Labour Migration and Development in Contemporary China

Since the mid-1980s, mass migration from the countryside to urban areas has been one of the most dramatic and noticeable changes in China. Labour migration has not only exerted a profound impact on China's economy, it has also had far-reaching consequences for its social development. This book examines labour migration in China, focusing on the social dimensions of this phenomenon, as well as on the economic aspects of the migration and development relationship. It provides in-depth coverage of pertinent topics which include the role of labour migration in poverty alleviation; the social costs of remittance and regional, gender and generational inequalities in their distribution; *hukou* reform and the inclusion of migrants in urban social security and medical insurance systems; the provision of schools for migrants' children; the provision of sexual health services to migrants; the housing conditions of migrants; the mobilization of women workers' social networks to improve labour protection; and the role of NGOs in providing social services for migrants. Throughout, it pays particular attention to policy implications, including the impact of the recent policy shift of the Chinese government, which has made social issues more central to national development policies, and has initiated policy reforms pertaining to migration.

Rachel Murphy is University Lecturer in the Sociology of China and Faculty Fellow at St Antony's College, University of Oxford. Her publications include *How Migrant Labor is Changing Rural China* (2002), *Chinese Citizenship: Views from the Margins* (2006, co-ed with V.L. Fong), *Media, Identity and Struggle in 21st Century China* (2008, co-ed with V.L. Fong) and articles in *China Quarterly, Population and Development Review* and *Journal of Peasant Studies*.

Comparative development and policy in Asia series

Series Editors
Ka Ho Mok (Faculty of Social Sciences, The University of Hong Kong, China)
Rachel Murphy (Oxford University, UK)
Yongjin Zhang (Centre for East Asian Studies, University of Bristol, UK)

Cultural Exclusion in China
State education, social mobility and cultural difference
Lin Yi

Labour Migration and Social Development in Contemporary China
Edited by *Rachel Murphy*

Labour Migration and Social Development in Contemporary China

Edited by
Rachel Murphy

Routledge
Taylor & Francis Group

LONDON AND NEW YORK

Transferred to digital printing 2010

First published 2009
by Routledge
2 Park Square, Milton Park, Abingdon, Oxon OX14 4RN

Simultaneously published in the USA and Canada
by Routledge
270 Madison Avenue, New York, NY 10016

*Routledge is an imprint of the Taylor & Francis Group, an Informa
business*

© 2009 Editorial selection and matter; Rachel Murphy; individual chapters
the contributors

Typeset in Times New Roman by Keyword Group Ltd

British Library Cataloguing in Publication Data
A catalogue record for this book is available from the British Library

Library of Congress Cataloguing in Publication Data
Labour migration and social development in contemporary China/edited
by Rachel Murphy
    p. cm. – (Comparative development and policy in Asia series ; 2)
    Includes bibliographical references and index.
    1. Migrant Labour–China. 2. Migration, Internal–Social
      Aspects–China. 3. Migration, Internal–Economic Aspects–China.
      4. China–Social policy. I. Murphy, Rachel, 1971–
    HD5856.C5L33 2008
    331.5′440951–dc22   2008012955

ISBN10: 0-415-46801-9 (hbk)
ISBN10: 0-415-59029-9(pbk)
ISBN10: 0-203-89059-0 (ebk)

ISBN13: 978-0-415-46801-5 (hbk)
ISBN13: 978-0-415-59029-7(pbk)
ISBN13: 978-0-203-89059-2 (ebk)

Contents

Tables

Contributors

Fang Cai Professor and Dean of Population Studies and Director of the Institute of Population and Labour Economics, Institute of Population and Labor Economics, Chinese Academy of Social Science, No.5 Jianguomennei Dajie, Beijing 100732, China

Jude Howell Professor and Director of the Centre for Civil Society, London School of Economics, Rugby Road, Brighton BN1 6EDE, Sussex

Caroline Hoy Principal Consultant, Hall Aitken Associates, Associate Research Fellow, Scottish Centre for Social Science Research in China, University of Glasgow, Hall Aitken, 3F 93 West George Street, Glasgow G2 1PBY, UK

Rachel Murphy University Lecturer in the Sociology of China, St Antony's College, 62 Woodstock Road, Oxford, OX2 6JF, UK

Pun Ngai Associate Professor of Social Science, Division of Social Science, Rm 3364, Academic Building, Hong Kong University of Science and Technology, Clear Water Bay, Hong Kong

Ran Tao Associate Professor in Development Economics, Centre for Chinese Agricultural Policy, Chinese Academy of Sciences, Centre for Chinese Agricultural Policy, Chinese Academy of Sciences, Room 3812, No, Jia 11, Datun Rd. Anwai, Beijing 100101, China

Dewen Wang Professor of Economics and Chief of Social Security Research Division, Institute of Population and Labor Economics, Chinese Academy of Social Sciences, CASS No.5 Jianguomennei Dajie, Beijing, China, 100732

Ya Ping Wang Reader in Urban Studies and Director of the Scottish Centre for Chinese Urban and Environmental Studies, School of the Built Environment, Heriot-Watt University, Edinburgh EH14 4AS, UK

Yanglin Wang Professor, College of Urban and Environmental Sciences, Executive Deputy Head of the Graduate School, Peking University, Haidian District, Beijing 100871, China

T.E. Woronov Assistant Professor in the Bureau of Applied Research in Anthropology, University of Arizona, Tucson, Arizona 85721, USA

Foreword

Dr Frank Laczko

Head of Research, International Organization for
Migration, Geneva.

At the international level, there is growing recognition that, if properly managed, migration can deliver major benefits in terms of development and poverty reduction. At the UN General Assembly High Level Dialogue on International Migration and Development, held in September 2006, and the Global Forum on Migration and Development, held in July 2007, there was a mounting recognition that migration holds considerable potential for economic and social development.

Though there has been increasing attention paid to the potential role migration can play in fostering development, most of that attention has tended to focus on international migration. Internal migration has been somewhat neglected but is also an extremely important policy area. Indeed most migration actually occurs within borders. China alone counted some 126 million internal migrants in 2004, at a time when the total number of international migrants for all countries was approximately 200 million.

In recent years, as Skeldon (2003) has noted, the word 'migration' has nearly always been associated with 'international migration', while internal migration has been subsumed under such terms as 'population distribution' or 'urbanization'. Those working on international migration seldom consider internal migration as relevant to their interests and vice versa.

This book is highly timely and of relevance to current policy debates about migration and development because it focuses on the relationship between *internal labour migration* and *development* in one of the countries of the world – China – where rapid internal migration has had substantial implications for development. China is also a very interesting case study because it is an example of a country that has decided to encourage and manage internal migration as part of its efforts to promote development. China's top leadership has referred several times to rural labour migrants as a major contributing factor to China's development.

Traditionally, migration has often been seen as a negative factor for development. Migration was viewed as the result of poverty and lack of development, or as a factor contributing to poverty in urban or rural areas. Internal migration, therefore, was sometimes considered as an obstacle to development that had to be restricted and controlled. This book does not ignore these problems, but it highlights the ways in which migration can contribute to development.

For instance, many internal labour migrants may work long hours for low pay and in poor working conditions. Their right to access the available health, education and social services may be tied to their rural residence and is not transferable to their new location in cities.

However, internal migration can also have a positive impact on development and poverty reduction. Internal migration has the potential to contribute to development in a number of ways. By supplementing their earnings through off-farm labour in urban areas, rural households diversify their sources of income and accumulate more collective capital. In the short term, migration may result in the loss of local financial and human capital, but it can also be beneficial and contribute to the long-term development of rural areas.

In particular, as outlined in detail in one of the chapters of this book, internal migrants' remittances can be larger than international remittance flows and therefore internal remittances can have an even more important impact on development. Internal remittances from urban employment can supplement rural incomes, boost consumption in rural areas, contribute to household savings and stimulate local economic development. In China, an official survey suggests that remittances are likely to become a more important source of income for rural households than earnings from agriculture (International Organization for Migration 2005).

Migrants who maintain links with their area of origin are more likely to transfer resources (remittances, investments, human capital and information) to their home base, and thereby help to raise the standard of living in rural communities and contribute to local economic development. Returned migrants, in particular, constitute an important potential source of investment, entrepreneurship and experience which can benefit the local population.

Many governments in the developing world have only recently begun to examine policies and strategies to enhance the positive linkages between internal migration and development in a systematic way (Laczko 2005). This book provides important new information and insights about internal migration in China. It provides us with a broader understanding of internal migration and development challenges in China and suggests way in which policy-makers across the world can learn from China's experience.

References

International Organization for Migration (2005) 'World Migration Report 2005', International Organization for Migration, Geneva

Laczko, F. (2005) 'Introduction: migration, development and poverty reduction in Asia', in *Migration, Development and Poverty Reduction in Asia: Internal Migration Lessons from Asia.* Geneva: International Organization for Migration

Skeldon, R. (2003) 'Inter-linkages between internal and international migration and development in the Asian region', paper presented at the Ad Hoc Expert Group Meeting on Migration and Development, organized by the United Nations Economic and Social Commission for Asia and the Pacific (ESCAP), Bangkok, 27–29 August, 2003

Acknowledgements

The contributors and editor express their thanks to Dr Frank Laczko, Director of the Research and Publications Department and the International Organisation for Migration, Geneva, for initiating and supporting this project on labour migration and social development in China. He commissioned the background papers on which most of the chapters in this volume are based, encouraged the production of a collection of policy-oriented papers on migration and social development linkages and provided invaluable comments and suggestions to the editor and authors. The chapters by Dewen Wang and Rachel Murphy in this volume were published in an earlier version in 2006 in the IOM Migration Research Series, and we are grateful for being able to present this material here. The contributors and editor also thank Ms Ilse Pinto-Dobernig of IOM for her tireless and meticulous copy-editing of the chapters in this volume. The editor also wishes to thank Dr Frank Pieke and Mr Daniel Holloway of the British Inter-University China Centre at the University of Oxford for support in bringing the book to production. Other important help with this volume came from the series co-editor Professor Ka Ho Mok, the Asian Studies editors at Routledge, Mr Tom Bates and Mr Peter Sowden, and the reviewer at Routledge.

Introduction

Labour migration and social development in China

Rachel Murphy

Introduction

In academic and policy discussions the social development aspects of labour migration in China tend to be eclipsed by more dominant economic concerns and paradigms. Migrants are often viewed as economic labourers rather than as complex social beings. Migrants are perceived to belong to a homogenous economic group or stratum, referred to in Chinese as *dagongzu*, rather than recognised as heterogeneous individuals who participate in various social relationships. Migration is seen as a low-cost way to generate remittances which alleviate inter- and intra-regional inequalities. Yet these costs are only low if counted in economic terms. The social costs are far greater and are more difficult to tally. Urbanisation too tends to be viewed through an economic lens of development and modernisation rather than as a process that necessitates social inclusion and the extension of social welfare services to new settlers. Relatedly, housing for migrants is seen mainly in terms of providing infrastructure for labourers rather than offering places in which human beings can maintain social relationships and feel security and emotional belonging. In this volume we aim to redress the relative neglect of social concerns by incorporating them into our exploration of salient aspects of the relationship between migration and development.

Social development and economic development are two sides of the same coin. Social development is an approach to social change which emphases the use of research, policies, and planned interventions to promote the wellbeing of the population as a whole in conjunction with the process of economic development (Midgley 1995). Managing social problems, meeting social needs and providing social opportunities through poverty alleviation, employment, livelihood security, education, healthcare and social integration enables individuals, families and communities to contribute to economic development. This in turn enable more people to help generate the institutional, human and economic resources that sustain social development (Birdsall 1993; Cook 2002; Midgley 1995).

Labour migration, a phenomenon integral to China's current rapid economic growth, is intimately interlinked with challenges to, and opportunities for, social development. Labour migration is responsive to regional and sectoral income inequalities; involves the exclusion of rural people from urban health, education, housing and social security systems (Li and Piachaud 2006); and draws on and reinforces various overlapping socio-economic divisions such as rural/urban, transient/resident, man/woman, educated/uneducated, employer/employee.

The factors which affect social development, including those related to migration, are difficult to delineate because any one aspect of a person's attributes or wellbeing may be affected by multiple cross-cutting factors, not all of which are observable. Failure to meet one kind of social need can severely constrain possibilities for meeting other kinds of social need (Alkire 2002). A person's health may be affected by their housing conditions, social integration, working conditions, gender, and access to medical insurance, healthcare and sexual health information. Their working conditions may in turn be affected by their education level, their capacity to organise to enforce labour regulations and the level of economic need in the rural family. Even in a situation of overall reduction in income poverty, non-income aspects of disadvantage arising from social and cultural norms may constrain people with particular attributes as they attempt to improve their wellbeing (Cook 2002). For instance, the adaptation of gender norms to the needs of capital accumulation may mean that in the new setting of the urban factory, employment does not 'empower' a rural woman but instead exposes her to new forms of risk and subordination which prevent her from pursuing objectives beneficial to her wellbeing such as education, social activities or better working conditions.

It is also difficult to delineate the inter-relationships between migration and development more broadly. First, development is both a cause and effect of migration. Development in the form of new roads and infrastructure both within and to areas of origin propels out-migration, while migration stimulates development in the form of urbanisation and return flows of resources and ideas to rural areas (Rhoda 1983). Second, migration simultaneously exerts positive and negative influences on development. Migration can reduce population pressure on land and resources in origin villages (Croll and Huang 1997; Galbraith 1980; Murphy 2002; Rempel and Lobdell 1978) and can contribute to economic growth in the cities (Nelson 1976; Parnwell 1993). At the same time it can result in the abandonment of farmland; the desertion of elderly people, spouses and children left behind in the rural areas (Croll and Huang 1997; Lipton 1980; Murphy 2002); the spread of diseases between rural and urban areas (de Lisle 2003); and the overcrowding of destination cities (DFID 2004; Nelson 1976; Parnwell 1993). Third, it is not always possible to discern all the factors that interact with migration or that operate independently of migration to produce particular outcomes – negative and positive trends in social and economic development may be only tangentially related to migration. Fourth, the outcomes of migration always depend on context. Here context refers to the social, economic, environmental and demographic conditions of both the origin and destination areas that shape each migration stream. It is therefore difficult to generalise about the developmental effects of migration and to make one-size-fits-all policy recommendations (de Haan 1999; Ellis 1998).

Finally, migration results in a reallocation of resources within households, communities and nations that inevitably produces winners and losers. Focusing on one group may produce verdicts that migration is positive for development whilst focusing on other groups may lead to the conclusion that migration is negative.

While keeping in mind the complexity of the inter-relationships between migration and social development in this book we explore a delineated set of pertinent questions: What is the impact of labour migration on China's poverty alleviation efforts in rural and urban areas? What are the implications of remittances for the wellbeing of the migrants who send their wages home and for the rural family members that receive the money? How do generation and gender affect the distribution of the social costs and economic benefits of migration within families? What is the best way to reform China's internal passport system (*hukou*) and to build a social and medical insurance system that includes migrants? How do migrant teachers, students and parents respond to the exclusion of their children from urban schools? What kinds of reproductive and sexual health services are available to migrants? Why is poor quality housing a common feature of migrants' urban lives? How do factory living and working conditions impact on workers' wellbeing? What are the possibilities for workers to organise and transform their own social networks into sources of support? How might NGOs improve the protection of and services for migrants?

Probing questions such as these is important for at least three reasons. First, vast numbers of people are involved in or immediately affected by migration – women, men, married people, singletons, parents, children, young, old, workers and farmers – and this has implications for their wellbeing. Second, over the past two decades, migration policy debate in China has undergone dramatic change from a focus on curbing to managing to actively encouraging migration. Both Chinese and internationally based academics and development practitioners have exerted some influence on the direction of this shift (Xiang and Tan 2005). Ongoing informed contributions to these deliberations may shape new directions in policy content and support innovations to confront the formidable obstacles associated with both implementing the new directives and ensuring that people benefit from them. Finally, unlike other countries with long migration histories, rural–urban migration appeared rather suddenly in China in the 1980s as a highly visible phenomenon. This explains in part why, unlike in some countries in South Asia and South East Asia, in China migration has become central to policy discussions about development. Analysing the successes, failings and outstanding problems of the Chinese case offers opportunities for reflection on policy approaches that may have wider relevance beyond China. This is valuable given that socio-economic systems are interconnected and the outcomes of regional internal migration are of direct relevance globally. Indeed the chapters in this volume began life as background migration policy papers which were commissioned by the Research and Publications Department of the International Organisation for Migration (IOM) in Geneva for precisely these reasons, each of which is considered in greater detail below.

The numbers

Understanding the relationship between migration and social development is valuable because the wellbeing of hundreds of millions of people is affected. Worldwide the vast majority of migrants move within their nation's boundaries. In 2000, the total figure for international migrants living in various parts of the developing and developed world was 175 million (United Nations Population Fund 2004). The same year, in China alone, an estimated 180 million rural migrants were working in urban areas. Even though the total number of internal migrants across the globe far exceeds that of international migrants, internal migration has received only a fraction of the attention of international migration in the media and scholarly literature. This neglect merits redressing because the immense scale of internal migration means that the critical links between migration and development are likely to have the greatest consequences within rather than between countries.

As already mentioned, in both origin and destination areas there are many more people other than the actual migrants themselves who participate in the migration process or who are affected by it. Much migration occurs as part of rural livelihood diversification (Ellis 1998). This means that households do not depend on farming alone for their livelihood but instead minimise their risks and raise their returns to available labour by incorporating different income sources into their household budget (Ellis 1998; Hare 1999). Family members remaining in the rural areas, for instance elderly people, often facilitate the migration of their adult children by caring for grandchildren and looking after the migrants' land. For their part, city-based migrants endure much hardship in order to send a high proportion of their wages home, and this often entails compromising their own health and enduring protracted separation from spouses and loved ones.

In China, large numbers of people will continue to participate in and/or be affected by rural to urban migration for the next two to three decades (Central News Agency 2005; Tao, this volume; Qu 2005; Zhou 2004). In 2003, some 23 per cent of total rural labourers had migrated, a further 28 per cent worked in rural enterprises, while the remaining 49 per cent (around 350 million) continued to rely on farming (Huang and Zhan 2005b).[1] Given that most of these rural labourers lack sufficient land and resources to sustain their livelihoods, they are potential migrants.

Policy shift

Understanding the relationship between migration and social development is also valuable because it can help us to interpret the consequences of major policy changes underway in China which affect the environment in which migrants live and work. These policy changes are best explained through a brief review of the particular history of migration in China.

Labour migration started to become a visible phenomenon in China during the early 1980s when a return to household farming and the emergence of private markets enabled rural people to move to the cities. At this time policy-makers became alarmed at the sudden appearance of rural people congregating at railway platforms, post-office counters, construction sites, the city outskirts, and on urban public transport. Policy-makers' initial response was to call for the flow of migrants to be curbed through the enforcement of the household registration (*hukou*) system. The *hukou*, established in 1958, has systematically allocated rights and public goods entitlements on the basis of place of residence and the occupational designations of 'agricultural' or 'non-agricultural'. Under this regime rural migrants in the cities without a temporary urban residence permit could find themselves vulnerable to detention, fines and repatriations. They were commonly denied access to many jobs reserved for the new urban unemployed. They also faced exclusion from public goods and services essential to social development such as healthcare and schooling for their children (Mallee 1995; Solinger 1999; Wang 2006; Wang and Zuo 1999; Zweig 1997).

By the early 1990s, policy discussions in China about migration centred largely on whether it was better to find ways to keep people on their farms or to encourage them to move to the cities. Ultimately, the government decided on the latter. It sanctioned the vision of a wholly urbanised and industrialised future (Li and Piachaud 2006). Migrants were no longer viewed as problems – rather they were just part of an inevitable historical modernisation process. From this perspective, minimisation of the negative effects and maximisation of the benefits of this process would require careful management by the government (Mallee 1995/1996; Murphy, 2002).

One obstacle to accelerated urbanisation, however, has been the circular nature of migration. In the absence of *hukou* reform, adequate urban wages, tolerable living and working conditions, job security, and the establishment of a national and unified social security system, migrants prefer to maintain a relationship with the land and the welfare security it offers. They retain close links with their family in the village and often intend to return to their homes (Murphy 2002). This so-called 'semi-proletarian' status of peasant workers depresses labour costs for urban factories because family members and the land in the village provide a security net for the migrant while the rural economy shoulders the costs of labour reproduction (Murphy 2002). The semi-proletarian character of migrants has at least three effects. First, the pressure on rural migrants to support family members who remain in the village reduces their ability to establish themselves in the city, by for example renting or buying decent quality housing (Wang and Wang, this volume). Second, low labour costs for migrants become ideologically legitimate because the status of 'peasant worker' prevents the migrants from recognising themselves or being recognised by either capitalists or urban planners as members of the urban working class who are entitled to urban benefits (Howell, this volume; Pun, this volume). Third, the 'semi-proletarian' status of migrants hinders a rural exodus and the formation of large-scale modernised mechanised farms. A persistent strand of policy debate therefore examines how to facilitate rural land consolidation while enabling urbanisation and migration without overwhelming urban infrastructure and public services or causing instability (Li 2006; Li and Piachaud 2006; Mallee 1995/1996; Tao and Xu 2007).

More recently a further policy shift has occurred. Policy-makers have moved on from an understanding that migration is inevitable and needs to be managed to stating that migration should be actively encouraged and that migrants are not just labourers, but are also individuals with needs and rights that must be protected. One indication of this policy shift was the State Council No. 2 Document (2002) which announced four principles for a reformed approach to migrants: fair treatment, reasonable guidance, improved management and better services for migrants (Huang and Zhan 2005a). This was followed in 2003 by the State Council No. 1 Document which called for:

- the abolition of unfair restrictions on rural labourers seeking work in urban areas;
- guarantees that migrants' wages are paid on time;
- improvements to the occupational health and safety environment of migrants;
- the provision of free legal and skills training for migrants;
- the provision of education for migrants' children of the same quality as that received by urban children;
- improvements in the administration of migrant populations with regard to family planning, children's schooling, healthcare and legal aid services (Huang and Zhan 2005a; Huang and Zhan 2005b).

Subsequently in March 2006 the State Council issued 'Directives on Matters of Migrant Workers', its first-ever comprehensive policy statement on migrants. This wide-ranging document identified the migrant workers' issue as being important for 'China's social and economic development situation as a whole'. Moreover it delineated a broad set of reform priorities aimed at improving migrants' living and working conditions, enhancing their legal and social protection, and promoting equality with urban residents (State Council Directives, 2006).

The policy shifts indicated in the above-mentioned policy documents were prompted by a convergence of changes in outlook across government and non-government sectors. First, the government recognised that migrants were contributing to national economic growth and that a productive economy reaps benefits from unfettered labour flows. Second, the Party state was concerned about the social unrest generated by widening socio-economic inequalities. Third, planners became aware that equality of access to health, education and social welfare for all city dwellers was necessary for long-term sustainable urbanisation and industrialisation (Li 2006; Li and Piachaud 2006; Woronov this volume). Fourth, the NGO sector, both indigenous and international, started to address the abuse and discrimination faced by migrants. At the same time, the Chinese leadership grew more responsive to international norms and increasingly concerned about its international image. Meanwhile, a reorientation in the leadership's agenda from a sole focus on economic growth to a broader development agenda that includes social justice promoted further efforts to address the plight of migrants.

A major aspect of the effort to encourage migration has centred on the question of how to reform the *hukou* system. As discussed in the chapters by Dewen Wang and Cai Fang and by Ran Tao in this volume, the question of how to reform the *hukou* system has persisted throughout the 1990s and early 2000s, and has gained impetus in recent years. During the early 1990s some urban settlements relaxed their ban on rural in-migration. Small cities granted urban residence permits to rural household registration holders in exchange for a substantial urban construction fee. New citizens had to prove that they had a secure form of employment and a house. By the late 1990s some selected larger provincial cities were also relaxing their immigration policies. Rules varied from city to city, but usually those migrants with better education and the ability to buy a house were granted a blue permit household residency – the equivalent of a 'green card' – which was, however, beyond the reach of most rural migrants (see also Solinger 1999). In 2000 some interior provincial cities such as Shijiazhuang and Zhengzhou removed obstacles to urban residency, requiring only that the migrants had a secure livelihood. Four years later, however, Zhengzhou municipal authorities felt that the urban infrastructure and resources would not be able to cope with large numbers of new settlers and reinstated *hukou* restrictions.

Further reform of the *hukou* system followed in the wake of a tragic incident. In 2002 a university student was beaten to death by police in a migrant detention centre. The perpetrators had mistaken the victim for a lowly rural migrant. The death of the student at the hands of the authorities alerted the urban educated and middle classes to the vulnerability and suffering that faced rural people living in the cities. There was an outcry among intellectuals, and many legal scholars declared that it was unconstitutional to detain people simply because they lacked a residence permit. The following year the State Council abolished the use of coercive custody and repatriation (Congressional-Executive Commission on China 2005).

More recently, in November 2005, plans were announced to remove the legal division between rural and urban residents across 11 provinces and to protect the rights of the rural migrants in the cities (International Herald Tribune, 2005). Yet at the same time that higher level authorities were urging a relaxation in mobility restrictions, municipal authorities commonly continued to use *hukou* to exclude migrants from a range of services such as new city co-operative healthcare programmes (Congressional-Executive Commission on China 2005), thereby avoiding the need to allocate their fiscal resources to social welfare for new settlers (Zhao and Li 2006). The urgent question of how to link *hukou* reform with a mechanism for funding all people's social security entitlements and public goods has consequently remained unsolved. As Ran Tao's chapter in this volume shows, there has been a gradual shift towards employer-based social insurance schemes. Yet migrants' high mobility and adverse material circumstances make them generally unwilling to contribute to such schemes. Moreover, most migrants work in the unprotected informal sector or else they are denied employment contracts, which automatically excludes them from such schemes. Indeed the trend towards labour informalisation has caused some scholars to question the assumption, derived from Western historical experience, that urbanisation and industrialisation help to lay the foundations for the welfare state and vice versa (Breman 2006). Additionally, as is clear in the chapters by T.E. Woronov, Caroline Hoy, and Ya Ping Wang and Yanglin Wang in this volume, even without the discrimination of the *hukou* system, simply being poor is a barrier for most migrants in trying to obtain affordable schooling, healthcare or housing.

Policy learning

A final reason that examining the interaction between migration and social development is valuable is because China's national policy approach as well as localised policy approaches present opportunities for policy learning and reflection. At the national level, the case of China is noteworthy for the explicit and substantive incorporation of migration into the majority of its key poverty alleviation and development plans. As mentioned, this differs from the situation in some countries in South Asia and South East Asia where policy makers refuse to talk about migration because they see it as an embarrassment, a symbol of exploitation and illustrative of the failure of development (International Organisation for Migration 2005). In the case of China, the policy focus on the link between migration and social development occurs at all levels of government. Indeed, as Lei Guang has argued, local officials commonly understand 'local development' to mean that they must follow key facets of the central government's national reform agenda for creating a market economy (Guang 2005). Hugely ambitious projects which aim to increase prosperity through the market economy are coordinated by the central government and executed locally. One example is the Sunshine Training Project which was initiated in 2003 to provide rural youth and intending migrants with short-term vocational skills training either free of charge or at low cost. The presence and form of these projects at the local level are determined in part by local geographic and socio-economic conditions. They are also determined by local political commitment to harness migration for achieving regional development; indeed, such projects can be a relatively low-cost way for governments in poorer localities to demonstrate to their superiors and constituents that they are committed to local development. As an example, in parts of Anhui province, county governments run training programmes which cultivate a 'local brand' of female worker with distinctive regional skills (such as domestic work, cooking or sewing) to enhance their 'value' and competitiveness in urban labour markets (Anhui's Suxian creates ..., 2004; Anhui's brand-name ..., 2006).

The approach of the Chinese government to development policy is also worthy of attention because recent policy shifts and new directives not directly related to migration have nevertheless changed the circumstances in which migrants and their families try to improve their livelihoods (Huang and Zhan 2005a). Recent developments have included the removal of farmers' tax burden by reforming the rural tax system; increasing investment in rural health, education, and public works (Zhang *et al.* 2007); and adopting measures to counter discrimination against girls and to promote gender equality more widely. So far the record in these areas is patchy but there is overall improvement. With regard to tax for fee reforms, even though in many rural localities, governments lack sufficient funds for providing public goods and are having to reinstate some unofficial fees, the total farmer's burdens has nevertheless decreased (Chen and Wu 2006; Kennedy 2007). With regard to gender equality, even though inequalities persist in many areas of social life, the gap between boys' and girls' rates of completion of compulsory education has been closing (Hannum and Adams, forthcoming). Cautious optimism about the deepening of a social justice agenda suggests that positive changes in rural areas may affect how migrants use their wages, how rural households use their remittances and the extent to which individuals are 'empowered' when participating in the migration process.

Yet despite improvements in support and protection, people from rural areas continue to be confronted by considerable obstacles in their pursuit of better life chances. One such obstacle is the erosion of collective welfare schemes and weaknesses in institutional capacity to provide social security. An important example is that of the healthcare system which has been privatised to the extent that both rich and poor lose out. The rich are over-prescribed drugs while the poor are unable to afford even basic treatment (Bloom and Fang 2003; Duckett 2007). Admittedly many social groups in China are disadvantaged by this situation. However migrants face particular health vulnerabilities which are exacerbated by the fact that they are less likely than urban residents to be covered by medical insurance and they face greater exposure to hazardous occupational environments (Xiang 2005).

A further factor hindering migrants' opportunities for social development is the tendency of Chinese policy-makers to devise services according to categorisations of 'personhood', a practice which causes certain people to be overlooked. This is especially relevant to poverty alleviation. The Ministry of Agriculture and the Office of Poverty Reduction and Development help the rural poor while the Ministry of Civil Affairs and the Ministry of Labour and Social Security help the urban poor. This rural/urban division leads to the neglect of city-based migrants who, lacking an urban *hukou*, are excluded from many sources of support available to urban residents while remaining ineligible for aid provided to the rural poor (Zhan 2005). A further example, discussed in Hoy's contribution to this volume, is that government agencies tend to provide sexual health services only to married people, thereby overlooking those who are sexually active outside marriage. Single migrants and/or their family members may therefore face particular disadvantage on account of ignorance on the part of health agencies and people's consequent exposure to certain kinds of risk.

There has however been growing enthusiasm within China for various forms of policy learning to address deficiencies in designing social programmes. China has a long-established tradition of applying policy learning across localities. This pattern of geographic transference is now being applied to migrants. For example, cities are copying good initiatives such as migrant night schools. A further element of policy-learning involves including NGOs in delivering social services to migrants, and learning from these entities about more effective ways to provide services. As an example, through engagement with NGOs, local officials may become familiar with concepts prevalent in 'development circles' such as 'gender', 'stigma', 'discrimination' (Zhan 2005) and 'community-based' approaches (Hoy, this volume), with NGOs in turn being influenced by the terminology and concepts of the international development community (Jacka 2006). Such familiarity may assist both local officials and NGO workers in designing policy interventions and support programmes which respond to peoples' needs more effectively, though may also impose ways of thinking and acting on indigenous NGO workers and their clients which are disconnected from local needs and interests (Jacka, 2006).

The chapters

Different dimensions of social development and their interactions with migration are evidently overlapping and cross-cutting. Each chapter in this book therefore necessarily engages with several interlinked social development themes, while key themes feature in more than one chapter. For instance, health appears in the chapters on poverty, remittances, social insurance systems, sexual health and housing; education is discussed in the chapters on poverty, remittances and migrants' schools; and gender is discussed in the chapters on remittances, sexual health and NGOs. While remaining mindful of the intersections between different social development themes, for reasons of clarity the chapters in this volume have been organised around three broad topics: livelihood security, public goods and the role of NGOs in organising to address the needs of migrants.

The essays by Wang Dewen and Cai Fang on poverty and by Rachel Murphy on remittances and rural livelihoods show that while the money sent home by the migrants has not turned the tide of rural–urban inequality, migration nevertheless exerts an equalising effect within rural areas by enabling the poor to obtain cash. Both essays demonstrate that by using this benefit, governments in poor rural regions have been able to integrate labour migration into their development planning. For instance, they have devised social development projects to enable households which lack credit, information and education to acquire the resources they need to send one household member to work in urban labour markets. Both essays also acknowledge that remittances raise rural incomes and improve living standards by facilitating increased spending on housing, health and education. Yet as Murphy cautions, the focus on remittances as a tool for promoting collective welfare obscures the human costs of migration and inequalities in their distribution.

The essays by Wang Dewen and Cai Fang and by Ran Tao examine the urban dimensions of the livelihood security challenges faced by migrants. Wang and Cai show that while rural–urban migration does not significantly increase income poverty in cities, labour market discrimination and social exclusion nevertheless expose rural migrants to considerable risks and vulnerabilities. Tao uses the *hukou* and social insurance systems to analyse the institutional underpinnings of these vulnerabilities. He argues that compared with the recent past, an urban *hukou* is no longer so closely associated with job-related social insurance, such as pensions, medical insurance and unemployment insurance. However, Tao also shows that the urban *hukou* remains closely tied with social assistance and other public services provided by city governments to permanent urban residents. He argues, therefore, that further policy changes to provide social security for migrant workers not only demand breakthroughs in China's *hukou* reform, but also a robust institutional social insurance framework appropriate for migrant workers.

The causes and effects of the exclusion of migrants from different kinds of public goods in the cities are explored in the essays by T.E. Woronov, Caroline Hoy, and Yaping Wang and Yanglin Wang. T.E. Woronov's chapter draws on her ethnographic field research conducted in a migrant-run school in Beijing to examine how straddling urban and rural social worlds and education systems affects the children's self-perception, hopes and life chances. Her ethnography provides a rich foundation for her assessment of the likely impact of new government policies pertaining to migrant children's education. On the one hand she is optimistic, conceding that recent policy trends at least recognise the needs and rights of migrant children. On the other hand, her optimism is tempered by concerns that the exclusion of migrant children from the urban mainstream is likely to continue. This is the result of significant latitude at the local government level in the extent to which migrant children's needs for education are met; the rapid changes in policies and variations in policies across cities and provinces; and the discrimination against migrants which remains deeply entrenched among urban planners and residents.

Caroline Hoy considers the difficulties faced by migrants in obtaining access to reproductive and sexual health services and education. She examines the reasons behind the failure of policy-makers, state service providers and NGOs to recognise that migrants are heterogeneous and that they have a sexual dimension to their lives. With regard to the state she suggests that measures be taken to move beyond the dichotomies of rural/urban, transient/permanent, married/single and male/female in programme design. With regard to NGOs she advocates an approach to outreach work that is free from moral censure.

Yaping Wang and Yanglin Wang draw on their survey data and ethnographic observation in Chongqing, Shenyang and Shenzhen to explore the hardships that many migrants face with regard to their living conditions. They show that the type of accommodation available to migrants varies according to the niches that they occupy in the urban labour market, but that regardless of whether they work in construction, in factories or in market trading, poor quality housing remains a common feature of their urban living experience. Wang and Wang suggest that for major improvements in migrants' housing conditions to occur, urban policy makers would need to shift their priorities (1) from collecting taxes and imposing management fees on landlords and (2) from carrying out urban regeneration and re-zoning projects to instead specifically focusing on the welfare of migrants.

The potential for migrants to organise and engage in collective action to improve their living and working conditions and to gain access to services is examined in the chapters by Pun Ngai and Jude Howell. Pun Ngai presents a case study of the Chinese Working Women Network. She documents how this organisation has inventively identified the relationships that women workers form through living collectively under the factories' dormitory regime as a resource that can be mobilised for mutual social support and for pressurising employers to comply with labour regulations. Jude Howell surveys the scope of activities of international and indigenous NGOs which target migrant workers in China. She finds that these entities have negotiated around the restrictions of a wary bureaucratic environment to organise and register themselves in a range of forms. Further, she documents how these entities have used the civic space that has emerged under the official banner of 'small government, large society' to offer an expanding range of services which contribute to migrants' wellbeing.

While acknowledging that the activities of NGOs are increasing in number and variety, however, the essays by Pun and Howell reveal at least four caveats. First, NGOs' engagement in direct advocacy on the part of migrant workers and the pursuit of a migrant empowerment agenda remain limited by unfamiliarity with these organisations on the part of China's government and society. Second, NGOs face difficulties when they attempt to move beyond service provision and engage in advocacy work. Indeed service provision is the state-endorsed remit of NGOs and quasi-NGOs in China (Cook 2002; Du 2004), yet it is often advocacy that enables more effective service provision. This is not to deny that in recent years some NGOs have engaged in advocacy work around migrants' rights (Jacka 2006; Pun, this volume), but the scope remains limited and their key task remains service provisioning. Third, in common with NGOs operating elsewhere in the world, NGO capacity in China is compromised by the absence of stable funding streams. Fourth, and also in common with the NGO sector elsewhere, there is much duplication of effort by different NGOs in some areas of outreach while other areas are overlooked (for other countries, see Edwards and Hulme 1996; Hulme and Edwards 1997).

Conclusion

Clearly, many migrants and their families have unmet social needs. This is in part because although the role of *hukou* as a tool for mobility restriction has declined, the institutional framework of *hukou* has continued to exist in a changing and increasingly marketised environment. In this environment local governments have responded to their increased control over local fiscal revenues and increased responsibility for providing public goods by continuing to use *hukou* to discriminate in determining peoples' entitlements to public goods (Zhao and Li 2006). At the same time, top-down policy mandates under the new social justice agenda have repeatedly exhorted municipal authorities to take seriously the needs of city-based migrants. So far, the responses of local governments to migrants' needs have been fragmented and much of the responsibility for social welfare has been shifted to employers. But, as Tao argues, such a fragmented response cannot adequately substitute for central state efforts to build social insurance systems which include migrants, the urban poor and the rural poor. This is because with a localised and fragmented approach and with increasing informalisation of the labour force too many people fall through the net.

Evidently most migrants cannot rely on state institutions to satisfy their social development needs. However, it is also difficult for migrants to achieve the socio-economic mobility necessary for them to obtain the resources to pay for their own needs. As Wang and Wang explain, migrants must make do in the face of low urban wages, ongoing obligations to support family members in the rural areas and institutional exclusion from mainstream affordable public goods and services. It is, moreover, difficult for migrant workers to organise as a collective entity to meet their own needs. Woronov demonstrates that migrants' prospects for using their own social networks to provide education for their children are constrained by ongoing discrimination against people of rural origins. Pun shows that rural women workers' possibilities for using their own social networks to create support groups are restricted by the wider political environment. And Howell explains that even though there is increased mobilisation around migrants' needs, migrants' voices are generally absent from projects organised *for* them. Meanwhile migrants face a unique set of constraints not encountered by other social groups in their efforts to organise *by* themselves.

Yet despite the formidable obstacles, gaps and failures with regard to meeting the social needs of migrants and their families, it is nevertheless the case that policy-makers and functionaries at different levels of the Party-state apparatus in China are making explicit, concerted and intensifying efforts to integrate labour migration into strategies for social and economic development. These efforts are evident in policy debates about the role of migration in national and regional social and economic development; in measures to enhance the capacity of the rural poor to migrate and obtain remittances; and in regulatory reforms to improve the labour protection, social care and housing safety of migrants living in the cities. There is now also increased, albeit cautious, discussion by state policy makers about a potential role for NGOs in helping to meet migrants' needs, though as Hoy and Howell caution, this cannot substitute for state commitment. Most encouragingly there is emerging recognition in Chinese policy-making circles that migration has human costs as well as the potential to generate economic resources; that migrants are not just economic labourers but are also social beings; that remittances cannot substitute for the investment of funds from other sources into social development projects; and further, that migration and urbanisation require sound social policies just as much as prudent economic regulation. By considering the impact of migration on livelihood security in rural and urban areas, the record of different levels of government in providing public goods and welfare, and the role of NGOs in meeting migrants' needs, the essays in this volume aim to encourage further discussions which link migration with the social as well as economic dimensions of development.

Acknowledgements

I gratefully acknowledge useful comments from Dewen Wang and Jude Howell on this chapter, and am particularly grateful for invaluable editorial and substantive feedback from Caroline Hoy.

Notes

1 Huang and Zhan estimate the numbers of surplus rural labourers who will migrate at 150 million. However, if 13 million migrate each year for the next 15 years then the total is closer to 200 million.

References

Alkire, S. (2002) 'Dimensions of human development', *World Development*, 30(2): 181–205.

'Anhui brand-name migrants call out to China's labour market' (Anhui mingpai mingong jiaoxiang zhongguo laowu shichang), *Xinhua Net*, 7 December, 2004. Available at: http://www.news.xinhua.com/fortune/2004-12/07/content_2304246.htm

'Anhui's Suxian creates a labour export brand name' (Anhui Suxian dazao laowu shuchu pinpai), *Xinhua Agricultural News Net* (*Anhui Nongye Xinxiwang*), 13 December, 2006. Available at: http://www.nmjyzx.com/news/readnews.asp?newsid=34004

Birdsall, N. (1993) 'Social development is economic development', Policy Research Working Papers, Policy Research Department, World Bank, Washington DC.

Bloom, G. and Fang, J. (2003) 'China's rural health system in a changing institutional context', IDS Working Paper 194, Institute for Development Studies, Sussex.

Breman, J. (2006) 'Informal sector employment', in Clark, D.A. (ed.) *The Elgar Companion to Development Studies.* Cheltenham: Edward Elgar, pp. 281–285.

Central News Agency (2005) 'China's population flow south-eastward may last 20 years', *The Epoch Times*, 28 February.

Chen G. and Wu C. (2006) *Will the Boat Sink the Water?: The Life of China's Peasants* (trans. Zhu Hong). New York: Public Affairs Press.

Congressional-Executive Commission on China (2005) 'China's household registration system: sustained reform needed to protect China's rural migrants', Congressional-Executive Commission on China, 7 October. Available at: http://www.cecc.gov

Cook, S. (2002) 'From rice bowl to safety net: insecurity and social protection during China's transition', *Development Policy Review* 20(5): 615–635.

Croll, E. and Huang P. (1997) 'Migration for and against agriculture in eight Chinese villages', *China Quarterly*, 149(March): 128–146.

de Haan, A. (1999) 'Livelihoods and poverty: the role of migration – a critical review of the migration literature', *Journal of Development Studies*, 36(2): 1–47.

de Lisle, J. (2003) 'SARS, Greater China and the pathologies of globalisation and transition', *Orbis*, 47(4): 587–604.

DFID (2004) 'China urban poverty study', Department for International Development, 31 October.

Du J. (2004) 'Gender and governance: the rise of new women's organizations', in Howell, J. (ed.) *Governance in China*. London: Rowan and Littlefield, pp. 172–192.

Duckett, J. (2007) 'Local governance, health finance, and changing patterns of inequality' in Shue, V. and Wong, C. (eds) *Paying for Progress in China: Public Finance, Human Welfare and Changing Patterns of Inequality.* New York: Routledge, pp. 46–68.

Edwards, M. and Hulme, D. (eds) (1996) *Beyond the Magic Bullet: NGO Performance and Accountability in the Post-Cold War World.* West Hartford: Kumarian Press.

Ellis, F. (1998) 'Household strategies and rural livelihood diversification', *Journal of Development Studies*, 35(1): 1–38.

Galbraith, J.K. (1980) *The Nature of Mass Poverty.* Harmondsworth: Penguin.

Guang, L. (2005) 'The state connection in China's rural–urban migration', *International Migration Review*, 39(2): 354–380.

Hannum, E. and Adams J. (forthcoming) 'Choices, hopes, and expectations: does gender still shape access to basic education in rural, northwest China?'.

Hare, D. (1999) 'Push and pull factors in migration outflows and returns: determinants of migration status and spell duration among China's rural population', *Journal of Development Studies* 35(3): 45–72.

Huang P. and Zhan S. (2005a) 'Internal migration in China: linking it to development' in *Migration, Development and Poverty Reduction in Asia* (ed.) International Organization for Migration. Geneva: IOM Research and Publications Department, pp. 67–84.

Huang P. and Zhan S. (2005b) 'Migrant worker remittances and rural development in China', paper presented at conference on *Migration and Development Within and Across Borders – Concepts, Methods and Policy Considerations in International and Internal Migration*, New York, 17–19 November, Social Science Research Council.

Hulme, D. and Edwards, M. (eds) (1997) *NGOs, States and Donors: Too Close for Comfort?* Basingstoke: Macmillan.

International Herald Tribune (2005) 'Rural migrants to get more rights in China', 2 November.

International Organization for Migration (2005) *Migration, Development and Poverty Reduction in Asia*. Geneva: IOM Research and Publications Department.

Jacka, T. (2006) *Rural Women in Urban China: Gender, Migration and Social Change*. Armonk: ME Sharpe.

Kennedy, J.J. (2007) 'From the tax-for-fee reform to the abolition of agricultural taxes: the impact on township governments in north-west China', *China Quarterly*, 189(March): 43–59.

Li, B. (2006) 'Floating population or urban citizens? Status, social provision and circumstances of rural-urban migrants in China', *Social Policy and Administration*, 40(2): 174–195.

Li, B. and Piachaud, D. (2006) 'Urbanization and social policy in China', *Asia-Pacific Development Journal*, 13(1): 1–26.

Lipton, M. (1980) 'Migration from rural areas in developing countries: The impact on rural productivity and income distribution', *World Development*, 8(1): 1–24.

Mallee, H. (1995) 'China's household registration system under reform', *Development and Change*, 26(1): 1–29..

Mallee, H. (1995/ 1996) 'In defence of migration: Recent Chinese studies on rural population mobility', *China Information*, X(3/4): 108–140.

Midgley, J. (1995) *Social Development: The Developmental Perspective in Social Welfare*. London: Sage.

Murphy, R. (2002) *How Migrant Labor is Changing Rural China*. New York: Cambridge University Press.

Nelson, J. (1976) 'Sojourners versus new urbanites: Causes and consequences of temporary versus permanent cityward migration in developing countries', *Economic Development and Cultural Change*, 24(4): 721–759.

Parnwell, M. (1993) *Population Movements and the Third World*. London: Routledge.

Qu H. (2005) *The Great Migration: How China's 200 Million Surplus Workers Will Change the Economy Forever*. London: HSBC Global Research.

Rempel, H. and Lobdell, R.A. (1978) 'The role of urban–rural remittances in rural development', *Journal of Development Studies*, 14(3): 324–332.

Rhoda, R. (1983) 'Rural development and urban migration: can we keep them down on the farm?' *International Migration Review*, 17(1): 34–64.

Solinger, D.J. (1999) *Contesting Citizenship in Urban China: Peasant Migrants, the State and the Logic of the Market*. Berkeley: University of California Press.

State Council (2006) 'The State Council Directives on Matters of Migrant Workers', State Council, Beijing.

State Council Office (2002) 'State Council Document No. 2: Improving Management and Services for Rural Labour Migrants, Beijing', summary presented in Huang Ping and Frank Pieke (2003) 'China migration country study', Regional Conference on Migration, Development and Pro-Poor Policy Choices in Asia, 22–24 June, 2003, Dhaka, Bangladesh, p. 36, available at: http://www.livelihoods.org

State Council Office (2003) 'State Council Document No. 1: Improving Management and Services for Rural Labour Migrants, Beijing', summary presented in Huang Ping and Frank Pieke (2003) 'China migration country study', Regional Conference on Migration, Development and Pro-Poor Policy Choices in Asia, 22–24 June, 2003, Dhaka, Bangladesh, p. 36, available at http://www.livelihoods.org

Tao, R. and Xu, Z. (2007) 'Urbanization, rural land system and social security for migrants in China', *Journal of Development Studies*, 43(7): 1301–1320.

United Nations Population Fund (2004) 'Meeting the challenges of migration: progress since the international conference on population and development'. UNFPA, New York. Available at: http://www.unfpa.org/publications/detail.cfm

Wang, F. and Zuo, X. (1999) 'Inside China's cities: Institutional barriers and opportunities for urban migrants', *American Economic Review, Papers and Proceedings*, 89(2): 276–280.

Wang, L. (2006) 'The urban Chinese educational system and the marginality of migrant children' in Fong, V.L. and Murphy, R. (eds) *Chinese Citizenship: Views from the Margins.* London: Routledge, pp. 27–40.

Xiang, B. (2005) 'An institutional approach towards migration and health in China,' in Jatrana, J. Toyota, M. and Yeoh, B. (eds) *Migration and Health in Asia* London: Routledge, pp. 161–176.

Xiang, B. and Tan, S. (2005) 'Does migration research matter? A review of migration research and its relation to policy since 1980s', Compas Working Paper No. 16.

Zhan S. (2005) 'Rural labor migration in China: Challenges for policies', Policy Papers No. 10, UNESCO, Paris.

Zhang, L., Luo R., Liu C. and Rozelle, S. (2007) 'Investing in rural China: Tracking China's commitment to modernization' in Shue, V. and Wong, C. (eds) *Paying for Progress in China: Public Finance, Human Welfare and Changing Patterns of Inequality.* London: Routledge, pp. 117–144.

Zhao, L. and Li, J. (2006) 'China's *hukou* system: Multifaceted changes and institutional causes', Discussion Paper 9 (June), China Policy Institute, University of Nottingham.

Zhou, T. (2004) 'Blueprint for China's economic growth', *China Daily*, 15 July.

Zweig, D. (1997) *Freeing China's Farmers: Rural Restructuring in the Reform Era.* London: ME Sharpe.

1 Migration and poverty alleviation in China

Dewen Wang and Fang Cai

Introduction

This essay examines the place of migration within China's ongoing poverty alleviation efforts. The discussion shows that rural–urban labour migration has played a pivotal role in poverty alleviation in rural areas, a role that has been enhanced through explicit government initiatives to link migration with the expansion of employment opportunities for rural people and to use education, training and credit assistance to enable individuals in very poor households to migrate. At the same time, the discussion shows that rural–urban labour migration presents new challenges for government approaches to poverty alleviation in cities where migrants face particular vulnerabilities on account of institutional and social exclusion. To provide context for delineating and evaluating the impacts of migration on poverty alleviation in China it is useful to begin with a brief overview of key poverty trends.

Since the launch of the reforms in the late 1970s rapid economic growth, together with a well-funded national poverty reduction programme, dramatically reduced the incidence of poverty in China's countryside. Official estimates indicate that between 1978 and 2005 the rural population living in poverty decreased from roughly 250 million to 23.7 million and poverty incidence fell from 30.7 per cent to 2.6 per cent over that same period (see Table 1.1).

The progress of rural poverty alleviation can be divided into four phases (see Table 1.1). The first phase was from 1978 to 1985. During this stage, the rural population living below the official poverty line fell from 30.7 per cent to 14.8 per cent (see Table 1.1). This 50 per cent reduction can be largely attributed to the success of the rural household responsibility system and the decollectivisation of agriculture with the attendant increase in agricultural productivity. The second phase of rural poverty alleviation started in 1986, but stagnated in the late 1980s and early 1990s. Although the Chinese government intentionally initiated large-scale regional development programmes to further reduce the numbers of remaining rural poor, the cooling down of economic growth stymied the pace of poverty reduction, and there were setbacks in 1989 and 1991 respectively. In 1993, the announcement of the '8-7' Poverty Reduction Plan marked the

Table 1.1 Rural poverty alleviation: 1978–2004

Year	Official poverty line			One-dollar-per-day criteria	
	Poverty line (yuan)	Poverty incidence (%)	Number of poor population (million)	Poverty incidence (%)	Number of poor population (million)
1978	100	30.7	250.0	—	—
1979	na	30.2	239.0	—	—
1980	130	26.8	220.0	—	—
1981	142	18.5	152.0	—	—
1982	164	17.5	145.0	—	—
1983	179	16.2	135.0	—	—
1984	200	15.1	128.0	—	—
1985	206	14.8	125.0	—	—
1986	213	15.5	131.0	—	—
1987	227	14.3	122.0	—	—
1988	236	11.1	96.0	—	—
1989	259	11.6	102.0	—	—
1990	300	9.4	85.0	31.3	280
1991	304	10.4	94.0	31.7	287
1992	317	8.8	80.0	30.1	274
1993	350	8.2	75.0	29.1	266
1994	400	7.7	70.0	25.9	237
1995	530	7.1	65.4	21.8	200
1996	580	6.3	58.0	15.0	138
1997	640	5.4	49.6	13.5	124
1998	635	4.6	42.1	11.5	106
1999	625	3.7	34.1	—	—
2000	625	3.4	32.1	—	—
2001	630	3.2	29.3	—	—
2002	627	3.0	28.3	—	—
2003	637	3.1	29.0	—	—
2004	668	2.8	26.1	—	—

Source: National Bureau of Statistics (2005, 2006); World Bank (2001).

beginning of the third phase. This plan called for national strategic action to reduce the numbers of rural poor by 80 million during the period 1994 to 2000. In implementing this programme, the government budgeted special poverty alleviation funds (PAFs) consisting of fiscal alleviation funds, food for work funds, and interest-subsidised loans to support economic growth in designated poor areas. With the accomplishment of this plan, the number of the rural poor dropped to 32 million with a poverty incidence of 3.4 per cent (China's News Office of the State Council, 2001).

Since the start of the new millennium, poverty alleviation in rural China has entered a new stage. The policy emphasis has been increasingly aimed at village-based and/or rural household-based development programmes rather than at previous county-based schemes. The new method endeavours to reach the remaining rural poor directly and to lift them out of poverty through improved targeting and financial utilisation.

Notwithstanding the remarkable progress already made, China is now facing a number of new difficulties in reducing poverty. First, the deceleration of poverty reduction in the countryside contrasts with the increasing marginal cost, which suggests greater difficulties in lifting the remaining rural poor out of poverty. The average annual change in poverty incidence dropped from 1.5 per cent in 1980s to 0.7 per cent by 1990s, falling further to 0.1 per cent since 2001. Meanwhile the annual Poverty Alleviation Fund input in 2001 and 2002 was 3.7 times higher than during the first half of 1990s, and double that of the second half of 1990s (see Table 1.1). Second, the nature of rural poverty have changed. The majority of the remaining rural poor are increasingly concentrated in remote and mountainous townships and villages in the western provinces; these people endure low educational attainment, poor health, bad living and reproductive conditions, and marginalisation (Asian Development Bank 2004; Cai and Du 2005). Their extreme and chronic poverty requires more specific poverty reduction measures.

Third, new issues of urban and migrant poverty have emerged. Prior to the 1990s, the poverty issue in urban China was of less significance than at present because the numbers of urban poor were far fewer, and the social and economic development needs of people were well provided for under the urban social relief system. For instance, in 1990 the number of the urban poor stood at 1.3 million with a poverty incidence of 0.4 per cent (World Bank 1992). Since the 1990s, the process of labour and social security reform in both state-owned enterprises and the urban private sector caused millions of workers to become redundant and tens of thousands of urban families to fall into poverty. Khan (1998) found that the urban poverty incidence increased by 12 per cent from 1988 to 1995. Updated statistics show that in 1999 the number of urban poor had reached 23 million with a poverty incidence of 5.1 per cent, with the poverty being more severe than in 1995 (Li 2001). If migrants are included, the issue of urban poverty becomes even more serious. Li (2001) reported that the poverty incidence of migrants is double that of urban residents with a local urban residence permit (*hukou*). According to a study of 31 large cities, the poverty incidence of migrants was over 50 per cent higher than for urban residents who had a local urban *hukou* and, in some cities, it was two to three times higher than for local residents (Hussain 2003). Therefore, attention to emerging urban poverty and to migrant poverty in particular, is an important component of China's future strategies for promoting the poverty alleviation, livelihood security and social harmony dimensions of social development.

Finally, income disparities between rural and urban areas and among regions have been worsening along with rapid economic growth. Rural–urban income inequality narrowed during the earlier years of reform, but has increased since the mid-1980s. From 1978 to 1985, the ratio of urban to rural per capita net income dropped from 2.57: 1 to 1.53:1, but then rose to 2.42:1 in 2004 (see Table 1.2). If we take into account the subsidised public services and welfare benefits in urban

Table 1.2 Income inequalities in China: 1978–2004

Year	Gini indices			Urban–rural income ratio	
	Rural	Urban	National	Current price	Constant price
1978	—	—	—	2.57	2.57
1980	24.99	na	na	2.50	2.35
1981	24.73	18.46	27.98	2.20	2.04
1982	24.40	16.27	25.91	1.95	1.79
1983	25.73	16.59	26.02	1.82	1.65
1984	26.69	17.79	26.89	1.83	1.63
1985	26.80	17.06	26.45	1.86	1.53
1986	28.48	20.66	29.20	2.12	1.69
1987	28.53	20.20	28.90	2.17	1.65
1988	29.71	21.08	29.50	2.17	1.51
1989	30.96	24.21	31.78	2.29	1.54
1990	29.87	23.42	31.55	2.20	1.64
1991	31.32	23.21	33.10	2.40	1.72
1992	32.03	24.18	34.24	2.58	1.78
1993	33.70	27.18	36.74	2.80	1.89
1994	34.00	29.22	37.60	2.86	1.95
1995	33.98	28.27	36.53	2.71	1.94
1996	32.98	28.52	35.05	2.51	1.85
1997	33.12	29.35	35.00	2.47	1.83
1998	33.07	29.94	35.37	2.51	1.86
1999	33.91	29.71	36.37	2.65	1.96
2000	35.75	31.86	38.49	2.79	2.04
2001	36.48	32.32	39.45	2.90	2.12
2002	na	32.65	na	3.11	2.30
2003	na	na	na	3.23	2.40
2004	36.92	na	na	3.21	2.42

Source: Gini indices are from Ravallion and Chen (2004). Rural urban income ratio comes from the National Bureau of Statistics (2005).

areas, the current rural–urban disparity in China would be the largest in the world (Li and Yue 2004). Ravallian and Chen (2004) documented that the overall rural and rural–urban income inequalities have been increasing since the beginning of 1980s. Labour market distortions are among the most important factors responsible for this increase, with significant direct and indirect effects on labour mobility and rural income.

The remainder of this chapter explores the relationship between migration and poverty alleviation. The first part examines the relationship between economic growth, employment and poverty alleviation; the second part describes the institutional conditions of migration, and the ways in which migration trends are shaped by trends in income inequality; the third part depicts the characteristics of poor households and analyses the contribution of migration to poverty reduction in sending places; the fourth part examines the urbanisation of poverty, and the final part concludes with policy suggestions.

Employment nexus between economic growth and poverty alleviation

Economic growth and poverty alleviation

Whether or not economic growth is pro-poor depends on its speed and quality. Since the start of the economic reforms, China's economy grew at an annual rate of about 10 per cent, with some cyclical characteristics over time. This rapid economic growth ensured China's large reduction of rural poverty, especially during the earlier reform period.

Plotting the poverty incidence against income growth reveals the importance of economic growth in the process of poverty reduction. Figure 1.1 demonstrates that the rural poverty incidence has declined along with the growth of per capita gross domestic product (GDP). Results from a simple regression model using rates of provincial rural poverty reduction from 1991 to 1996 on GDP growth reinforce the conclusion that provinces with more rapid per capita GDP growth also show a more rapid decline in the numbers of rural poor (World Bank 2001). The coastal provinces took the lead in initiating economic reforms and achieved faster economic growth, leading also to much faster rural poverty reduction compared to central and western regions.

Huang *et al.* (2005) regressed the national (and provincial) rates of rural poverty against per capita GDP and confirmed that economic growth significantly affects poverty reduction, though the growth elasticity of poverty declines with the increase in per capita GDP. They pointed out that if the inter-country data from the 2003 Global Development Report were used in the regression, a U-shaped relationship can be found between income level and poverty incidence with a turning point at US$ 25,000. Because all developing countries are below that

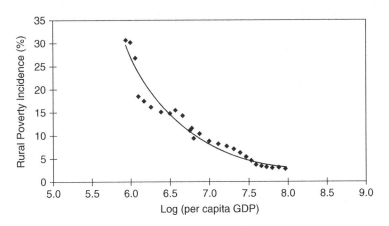

Figure 1.1 Economic growth and rural poverty alleviation.

Source: National Bureau of Statistics (2005), China Yearbook of Rural Household Survey (2005).

level, faster economic growth will have a stronger impact on poverty reduction during the initial stages of economic growth, but with a diminishing effect as an economy becomes wealthy and mature.

The above empirical evidence is consistent with findings from other sources (Chen and Wang 2001; Jalan and Ravallion 1998; Khan 2000), which testify that economic growth is an important factor in China's poverty reduction efforts, but that its effects have been declining since the mid-1980s. Slow agricultural growth is one factor that weakens the effects of economic growth on poverty reduction.

Between 1978 and 1984 agricultural growth in China was impressive. The value of agricultural output grew at an annual rate of 6.9 per cent, up from 2.5 per cent in the period 1952–1978. The annual growth rates for grain, cotton and oil seeds were 4.8, 17.7 and 13.8 per cent, respectively (National Bureau Statistics 2005). As a result, rural per capita income almost tripled and the number of rural poor was halved during that period.

Compared with the fast growth during the initial reform stage, agricultural growth slowed down to between 3.1 and 4.6 per cent between the late 1980s and the beginning of this century. In the meantime, the share of agricultural output in GDP decreased from 31.4 per cent to 15.2 per cent (see Table 1.3). Slow agricultural growth not only allows non-agricultural sectors to become the major contributors of economic growth, but also delays lifting the rural poor out of poverty because they rely mainly on agriculture for their household income and cannot equally enjoy the gains from the rapid growth of non-agricultural sectors. Chen and Wang (2001) used household survey data from 1990 to 1999 to empirically show that the poor have gained far less from economic growth than the rich, and that only 20 per cent of the richest had income growth equivalent to, or greater than, GDP growth. As a result, rising income inequality disconnects poverty alleviation from economic growth.

The increasing rate of accumulation and investment, which caused the difference between economic growth and income growth, is also one of the determinants of the

Table 1.3 Economic growth in China: 1978–2004

Time period	1978–1984	1985–1989	1990–1994	1995–1999	2000–2004
Growth rates (%)					
GDP	9.6	9.9	10.7	8.8	8.6
Agriculture	6.9	3.1	4.6	4.0	3.4
Industry	9.9	12.2	15.3	10.7	10.3
Tertiary	12.2	12.7	8.8	8.3	8.3
Per capita GDP	8.2	8.2	9.3	7.7	7.8
Composition of GDP (%)					
GDP	100.0	100.0	100.0	100.0	100.0
Agriculture	31.4	26.6	22.7	19.2	15.4
Industry	46.2	43.6	44.6	49.4	51.2
Tertiary	22.4	29.8	32.7	31.4	33.4

Source: National Bureau of Statistics (2005).

slowdown in poverty reduction (Khan 2000). The arithmetic average growth rates of per capita GDP, rural and urban per capita income from 1978 to 1984 are 8.2, 15.9 and 6.6 per cent, respectively; the growth of per capita GDP does not change from 1985 to 2002, while the growth rates of rural and urban per capita income drop to 4.3 per cent and 6.3 per cent, respectively. The growing gap between economic growth and income growth accounts significantly for a decrease in the speed of poverty reduction and an increase in the numbers of the urban poor (Lin and Li 2005). In fact, the high-speed economic growth propelled by investment comes to some degree at the expense of the slow growth of employment because it dilutes the 'trickle-down' effect.

Non-agricultural employment and poverty alleviation

Employment is the major activity for generating family income. The growth of rural income can be divided into agricultural and non-agricultural revenues. With slow agricultural growth, non-agricultural revenue becomes the main source of household income growth through non-agricultural employment. The share of wage income in rural household income increased from 17 per cent in 1985 to 34 per cent in 2004. In urban household income, wage income accounts for more than 70 per cent (National Bureau of Statistics 2005). If a family member is under-employed, the whole family income will be dramatically reduced. Therefore, wage-earning employment will be the crucial channel for maintaining family income and benefits from rapid economic growth.

Like fast agricultural growth, the rapid development of Township and Village Enterprises (hereafter referred to as TVEs) has also had a very positive impact on poverty reduction through the creation of non-agricultural employment. In 1978, total industrial production of TVEs was 49.3 billion *yuan*, accounting for 11.6 per cent of GDP. In 1992, this figure rose to 2,036 billion *yuan*, accounting for 38.6 per cent of gross national industrial product. From 1978 to 2003 the real growth rate of gross output was 28.0 per cent per year, creating millions of non-agricultural employment opportunities that facilitated the transfer of surplus rural labour. From 1978 to 2003 the number of people employed in TVEs rose from 28.3 million, accounting for 9.2 per cent of rural employment, to 138.7 million or 28.5 per cent of rural employment, with an average annual growth rate of 6.1 per cent (Ministry of Agriculture 2004).

The development of rural industrialisation has not been uniform across regions. In the early 1980s, the number of non-agricultural workers actually decreased in the poor central and western regions as the commune system was dismantled. In 2004, 53.4 per cent of TVE employment was concentrated in the eastern regions, compared to 27.7 per cent in the central and 19.0 per cent in the western regions (Ministry of Agriculture 2004). Regional differences in rural industrialisation caused regional differences in non-agricultural employment, thereby affecting the speed of rural poverty reduction. Since the rural poor are increasingly concentrated in remote and mountainous areas, slow agricultural growth and less developed industrialisation has limited poverty reduction in those areas.

The decline in the elasticity of employment that equals employment growth caused by corresponding GDP (or output) growth, further illustrates the decreasing effects of economic growth on employment and poverty reduction. As shown in Figure 1.2, the employment elasticity in non-agricultural sectors (including industry and tertiary industry) has a downward trend. In 1980s, China's annual average GDP growth was 9.8 per cent, and employment elasticity 0.56 per cent. In the 1990s, China's GDP grew at an annual rate of 9.3 per cent, while the elasticity of employment was 0.33 (National Bureau of Statistics 2005). TVE employment growth was strongest in the late 1980s and early 1990s, but its employment elasticity has also declined since then. The distortion of factors of production and economic restructuring towards capitalisation are the main reasons for the declining employment elasticity in non-agricultural sectors, which not only limits the full utilisation of China's abundant labour resources, but also hinders rural labourers from taking advantage of the opportunities of rapid economic growth to improve their quality of life.

Institutional reform and migration

Trends in rural to urban migration

In a country with such a huge population and so little land, rural labourers have strong incentives to migrate to the cities.[1] During the early 1950s, mobility into and out of the cities was relatively unrestricted and, in fact, a large number of rural labourers moved into the cities at that time. By the mid-1950s, however, the establishment of *hukou* segregated rural from urban areas and imposed strict controls on migration between them and across regions.

Prior to the start of the reforms, rural labourers were forced to work in agricultural sectors. In 1978, there were 285 million agricultural labourers, accounting for 70.9 per cent of the total labour force and 92.9 per cent of the total rural labour force (National Bureau of Statistics 2005). With the unfolding of rural and urban reforms, industrialisation cum urbanisation gradually transferred more and more rural labourers into non-agricultural sectors and urban areas.

In the early 1980s, when agricultural reforms first took place, few rural labourers migrated to other areas to work (see Table 1.4). Most of the two million rural migrants were craftsmen such as carpenters, construction workers and street vendors who moved between villages. With the improvement of agricultural productivity and the relaxation of government controls on rural–urban migration, increasing numbers of rural labourers migrated. By the end of the 1980s, China had a total of 30 million rural migrants.

Deng Xiaoping's visit to South China in 1992 ushered in a new round of rapid economic growth. The expansion of the private economic sectors drew large numbers of rural labourers out of agriculture, triggering the first large-scale migration wave across regions. The numbers of people employed in urban private enterprises or self-employed totalled 11.16 million in 1993, an increase

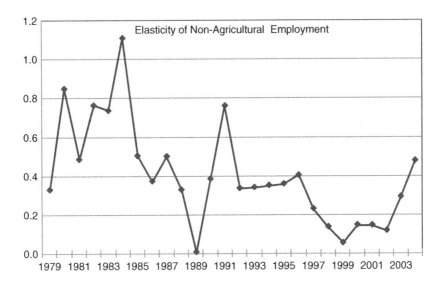

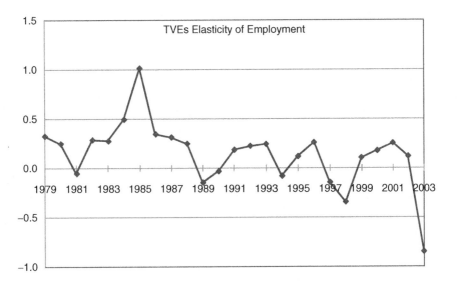

Figure 1.2 Employment elasticities in (a) non-agricultural sectors and (b) TVEs.
Notes: employment in non-agricultural sectors includes those in industry and tertiary industry.

Source: National Bureau of Statistics (2005).

Table 1.4 Number of rural–urban migrants: 1982–2004

Year	Out of township (million)	Average annual transfer (10,000s)
1982	2.0	50
1989	30.0	400
1993	62.0	800
1995	70.0	400
1996	72.2	223
1997	77.2	499
2001	89.6	348
2002	94.3	469
2003	98.2	390
2004	102.6	440

Source: Chen, X. and Zhang, H. (2005).

of 33.2 per cent over 1992. This figure reached 34.67 million in 1999 with an average annual increase of 3.92 million. In 1992 the number of projects receiving direct investment from foreign companies and from companies in the special administrative areas of Hong Kong, Macao and the Chinese Province of Taiwan had reached 48,764, 2.8 times that of 1991. The amount of actual foreign direct investment (FDI) reached US$ 11 billion, up 1.5 times from 1991, and kept growing in subsequent years (National Bureau of Statistics 2005). Rapid development of TVEs, especially in coastal areas, as well as in the booming economic development zones generated strong demand for cheap rural labourers.

The numbers of rural migrants doubled within four years. In 1993, rural migrants totalled 62 million, 22 million of whom had migrated across provinces, 2.07 and 3.14 times the respective figures for 1989. Subsequently, the numbers of rural migrants increased steadily to 70 million in 1994 and 75 million in 1995; of these, 25–28 million were inter-provincial migrants (see Table 1.4).

The 1997 South East Asian financial crisis had a negative impact on economic growth in Asia and the world at large. Consequently, export and TVE development in China suffered. The reform of state-owned enterprises and the urban employment system from the mid-1990s onwards led to massive layoffs of urban workers. The unemployment rate rose sharply and the job market shrank. Rural to urban migration slowed down to an annual average of 3.6 million.

Since 2001, rural to urban migration has again accelerated. From 2001 to 2004, the number of rural migrants increased to 4.12 million per annum. In 2004, rural migrants exceeded 100 million for the first time, accounting for 20.6 per cent of the total number of rural labourers. Despite the increasing migration of rural labourers, the expansion of non-agricultural sectors in the wake of China's accession to the World Trade Organization (WTO) has increased the demand for rural labourers. The eastern costal areas have experienced shortages of rural migrant workers since late 2002, causing local and structural labour market issues to emerge (Wang *et al.* 2005).

Changes in the institutional and policy environment for migration

China's rural to urban migration has evolved along with gradual institutional changes that have enabled the development of labour markets and the abolition of structural obstacles to mobility. Apart from the characteristics common to developing countries, China's rural to urban migration has some unique features associated with institutional transition. Specifically, the policy measures governing migration have gone through three stages since the beginning of the reform in 1978: permitting rural labour mobility, guiding rural labour mobility and encouraging rural labour mobility. Under the principle of market-oriented reform, the intention and focus of related policies have changed according to the macro-economic situation, leading to fluctuations in the numbers of rural migrants.

1980s: permitting rural labour mobility

The household responsibility system (HRS) initiated in the late 1970s made farm households the residual claimants of their marginal effort, and this stimulated an increase in farm productivity and released surplus labourers from agriculture. At the same time, the rapid development of TVEs, especially in eastern regions, increased the demand for rural labourers. As the focus of economic reform shifted from rural to urban areas, systemic reform gathered pace in urban areas. The tertiary sector in urban areas began to open up to rural migrants, creating more opportunities for labour mobility from agricultural to non-agricultural sectors and from rural to urban areas. In this situation, permitting rural labour mobility was not only what farmers desired, but was also a prerequisite for achieving urban economic growth.

In 1984, China began to allow farmers and agricultural collectives to engage in long-distance transportation and sale of 'three categories of agricultural and non-staple food',[2] as well as foodstuffs not included in the quota assigned by the state. This was the first time that Chinese farmers had the right to do business outside their hometowns. Farmers were also encouraged by the state to work in nearby small towns where emerging TVEs demanded labour. In 1985, the Ministry of Public Security promulgated 'Temporary Rules on Migratory Population in Cities and Towns', which required all those aged 16 and above who stayed in cities and towns for more than three months to apply for temporary residence permits. This policy provided the legal basis for charging a temporary residence permit fee and, to some extent, discouraged rural to urban migration.

From 1988 to 1990, to control the inflation induced by the overheated economy, the Chinese government adopted many economic measures, including reducing infrastructure investment as well as strengthening controls on financial markets, taxation and credit. As a result, many construction projects were suspended or stopped, and economic growth slowed down significantly. In order to protect urban workers, many rural migrants were fired, giving rise to a reverse flow of urban to rural migration. The development of non-agricultural sectors in rural areas also suffered greatly. The capacity of TVEs to absorb workers declined over the course of two consecutive years. According to the 'Notice on Employment Work', issued by the State Council on April 27, 1990, the government encouraged rural labourers to 'leave the land without leaving the township' and to seek employment locally. Stronger control was imposed on rural migrants in cities. Non-planned rural workers were let go and asked to return to the countryside.

The above-mentioned policies had the effect of deterring labour mobility. The population of rural migrants fell dramatically between 1988 and 1989, dropping by around one-third in several big cities. This decline was short-lived, however. By 1990–1991, in most cities the number of rural migrants had again reached the levels of 1988, and some even exceeded their previous levels.

1990s: guiding rural labour mobility

Growing income inequalities, the pressure of employment in agricultural sectors and the reform of the urban *hukou* system created push–pull forces leading to large-scale migration. In these circumstances, the former policies that suppressed labour mobility were obviously ineffective. Therefore, strengthening the management of labour mobility through the provision of employment guidance and employment services clearly emerged as the better option.

In January 1991, the Ministry of Labour and Social Security, the Ministry of Agriculture and the State Council Development Research Centre jointly decided to set up a pilot project for the promotion of China's rural labour employment. The implementation of this project proceeded in two phases. During phase I (1991–1994), the project was implemented in 50 counties, and during phase II (1994–1996), in eight provinces. Experiments were conducted to promote the non-agricultural employment of rural labourers in those counties and provinces to gain experience for expansion.

Based on the experience of the first stage of the pilot project, the Ministry of Labour and Social Security promulgated 'Temporary Rules on Managing the Employment of Rural Labour Migrating Across Provinces'. According to these, before leaving home, rural migrants were required to bring their ID card and other necessary documents to register at the local employment agency and obtain an employment registration card. After arriving at their destination, rural migrants needed to obtain the employment registration certificate for incoming migrant workers. Employment certificates for rural migrants (employment registration card plus employment registration certificate) served as a valid ID card for rural migrants and enabled them to enjoy the employment services provided by career centres. In November 1997, the State Council issued suggestions on the establishment of a comprehensive labour market planning and information service system for the creation of a labour market system.

The 'Procedure to Apply for a Temporary Living Card', promulgated by Ministry of Public Security on June 2, 1995, established rules for the use, effectiveness and change of the temporary residence card. Persons aged 16 and above who left their place of normal residence for more than one month for purposes other than visiting friends or family, travelling, seeking medical treatment or a business trip, were required to apply for a temporary residence card while waiting for a temporary *hukou*, valid for a maximum of one year.

In 1997, the *hukou* system was further relaxed. Small cities and towns began to grant *hukou* to rural migrants who either bought or built a house. In July 1998, the State Council approved 'Opinions on Solving Top Issues in Hukou Management', stating that migrants who had lived for a certain time in a city were permitted to obtain a local urban *hukou*, as long as they had a fixed residence, a stable and legal occupation or source of income. In the same year, the Ministry of Public Security issued new regulations relaxing the control over *hukou* registration, allowing persons who joined their parents, spouses and children in cities to register with an urban *hukou*. The reforms in urban welfare provision, such as the removal of rationing, the creation of a housing market, the adoption of more flexible employment policies and attempts to establish a social security system, have made it easier for rural labour migrants to make a living in cities.

Guiding rural labour mobility was the top policy priority in this period. Since the mid-1990s, however, after a large number of workers in state-owned enterprises (SOEs) were laid off, unemployed urban workers entered the urban labour markets and competed with migrants. In order to solve the unemployment problem of laid-off SOE workers, many cities adopted protective measures to exclude rural migrants from certain jobs. This policy had very limited effect and was rapidly abandoned in most cities, but continued to exist in some with high unemployment rates, such as Shengyang, to save jobs for urban workers (Solinger 1999) or for the arrangement of local re-employment after some TVEs collapsed in Jiangsu.

Since 2000: encouraging rural labour mobility

There have been a number of positive changes concerning the employment of rural migrants since 2000. The contribution of rural migrants to urban social and economic development has gradually been recognised by society, and urban residents have begun to change their attitudes towards migrant workers.

In January 2002, the State Council first released the 16-word policy of 'Fair Treatment, Good Guidance, Improving Management and Improving Services'. In January 2003, the State Council issued the 'Notice on How to Better Manage and Provide Services for Rural Migrants', requiring local governments to make greater efforts to provide better public employment management and services for rural migrants, eliminate unreasonable limitations on rural–urban migration, solve salary arrears and cuts, improve living and working conditions, provide more training opportunities, ensure schooling for the children of migrants and enhance management. In the first document of 2004, the central government pointed out that 'rural migrant workers have become a crucial component of the industrial workforce, and create wealth for cities and generate tax revenues'.

Under the new concept of 'fair treatment', unreasonable limitations on rural migrants have been gradually removed and mechanisms conducive to the employment of rural migrants have been put in place. In order to facilitate labour mobility and social stability, the State Planning and Development Committee, together with the Ministry of Finance, issued a notice in November 2001 requiring local governments to abolish all manner of fees levied on rural migrants, including temporary residence fees and management fees, family planning fees, urban expansion fees, labour adjustment fees, management and service fees, and construction enterprises' management fees. According to the 2003–2010 Nationwide Training Plan for Rural Migrants, jointly developed by the Ministry of Agriculture, the Ministry of Labour and Social Security, the Ministry of Education, the Ministry of Technology, the Ministry of Construction and the Ministry of Finance in September 2003, special funds have been secured by the central and local governments to provide training for rural migrants. The implementation of these policy measures will greatly improve the employment conditions of rural migrant labourers.

China's rural–urban labour migration began with the implementation of the household responsibility system, and reached its peak with the reform of the *hukou* system. Large-scale rural–urban migration has contributed towards China's transition to a market economy by providing the necessary labour. At present, the allocation of labour and other factors of production has been transformed in line with a market-economy approach. Although there has been progress in reforming the *hukou* system, much remains to be done, since the *hukou* system still associates employment with individual identity, and hinders labour mobility and social and economic development.

Migration as a labour market response to income inequality

In the 1980s, most rural migrants opted to 'leave the farm land without leaving the village', and chose to work in local TVEs. Since the early 1990s, fundamental changes in the pattern of supply and demand for agricultural products intensified the pressures for surplus rural labourers to transfer out of agriculture. In the meantime, booming economies in the southeastern coastal areas, such as Guangdong and Fujian, experienced increasing demand for cheap labour. Thus, the spatial imbalance between labour supply and demand triggered the first wave of migration. In 1993, the number of rural migrants was estimated at about 60 million, with one-third entering the cities. According to the National Bureau of Statistics Rural Survey, by 2004 this figure had risen to 120 million, half of whom were inter-provincial migrants.

Increasing regional mobility is the response of surplus rural labourers to growing income inequalities. Since the mid-1980s, when income inequalities started to widen, economic reforms gradually allowed market forces to play a greater role in resource and factor allocation. Benefiting from early openness, coastal provinces have been leading in both economic growth and the development of markets for factor allocation (Cai and Wang 2005), thereby eliminating the institutional obstacles that prevented factors of production from moving across regions, and creating the conditions to receive the massive inflow of rural migrants. Cheap rural migrants have, in turn, played an important role in driving economic growth in these regions.

The integration into international markets has accelerated the adjustment of economic restructuring in coastal regions towards labour-intensive industries, which utilise China's comparative advantage of abundant labour resources. In 2004, 92.6 per cent of the total export value in China was generated in the eastern region, against only 4.2 per cent in the central and 3.2 per cent in the western regions. In 2003, 85.7 per cent of foreign direct investment was invested in the eastern regions, against 11.0 per cent in central and 3.3 per cent in western regions (National Bureau of Statistics 2005). As a result, the main direction of migration is from the centre and western regions to the east.

The spatial distribution of migration reflects its responsiveness to income inequalities and institutional environments. By summarising data from a population survey and the 1900 and 2000 national censuses, Table 1.5 shows changes in spatial patterns of migration. In the period 1987–2000, intra-regional (mainly intra-provincial) migration dominated, with some changes occurring over time. As the share of inter-provincial migration within the eastern region increased, that within central and western regions declined. In the meantime, the share of migration between central and western regions decreased, while inter-regional migration from central and western to eastern regions increased. If we decompose the distribution of migration in 2000 into four types of migration (i.e. urban–urban,

Table 1.5 Regional distribution of inter-provincial migrants (%)

Destination	Origin			
	East	*Central*	*West*	*National*
East				
1987	49.7	61.7	44.2	52.0
1990	56.0	59.0	49.3	54.6
1995	63.5	71.8	56.5	63.1
2000	64.4	84.3	68.3	75.0
Central				
1987	31.3	21.8	21.2	24.6
1990	28.4	23.5	20.4	24.0
1995	20.5	12.7	13.4	18.8
2000	19.7	7.1	7.9	9.8
West				
1987	18.9	16.6	34.6	23.3
1990	15.6	17.5	30.3	21.4
1995	16.1	15.5	30.2	18.1
2000	15.9	8.6	23.9	15.3

Source: Cai, F. and Wang, D. (2003).

Notes: Migrants in 1987 refer to those who migrated between cities, towns and counties and stayed at destinations for six months or longer; migrants in 1990 refer to those who migrated between cities and counties and stayed at destinations for one year or longer; migrants in 1995 refer to those who migrated between counties, districts and counties and stayed at destinations for six months or longer; migrants in 2000 refer to those who migrated between townships, towns (*Zhen*) and communities (*Jiedao*), and stayed at destinations for six months or longer.

urban–rural, rural–urban and rural–rural migration), rural–urban migration accounts for the major part of the total, 40.7 per cent to be exact. Urban–urban migration ranks second, accounting for 37.2 per cent. Thus, these two are the main forms of migration in China. For reference, rural–rural migration accounts for 18.2 per cent of total migration, while urban–rural migration accounts for only 4 per cent of the total.

New surveys illustrate that rural migration is further concentrated in eastern regions. For example, in 2004 the five provinces with the highest ratio of migrants to provincial rural labourers, were Anhui, Jiangxi, Hubei, Chongqing and Sichuan, all experiencing a ratio of over 30 per cent and all located in central and western regions. More than 10 million rural labourers left Henan and Sichuan. Around 28.1 per cent of rural migrants chose to work in mega-cities and provincial capitals, 34.3 per cent in prefectures and less than 40 per cent in county towns and townships. The share of rural migrants who chose to work in the eastern regions increased from 64.3 per cent in 2000, to 68 per cent in 2003 and 70 per cent in 2004. Although the elasticity of migration to income disparity rose from 0.197 in the period 1985–1990 to 0.595 for the period 1995–2000, such an increase in mobility has not reduced income inequality mainly because of the unfinished reform of the *hukou* system and other factors that continue to accentuate regional disparity (Lin *et al.* 2004).

National development plans for poverty reduction through migration

In order to accelerate the pace of rural poverty alleviation, the Chinese government began incorporating migration into its national development plan. In 1986, the government defined the national criteria for designating rural poverty for the first time, and 331 counties with an annual per capita rural income of less than 150 *yuan* were identified as national poverty counties. During the late 1980s and the early 1990s, the Chinese government adopted a series of policy measures to strengthen the work of rural poverty alleviation in these counties, such as establishing specific leading agencies, providing funds and prioritising policies, and introducing an integrated regional development strategy to replace the traditional focus on providing social relief to needy households. Anti-poverty efforts also included attention to encouraging the transfer of rural surplus labourers into non-agricultural sectors as part of promoting integrated regional development. This mix of anti-poverty measures in these areas helped to bring about a rise in per capita rural income from 206 *yuan* in 1986 to 483.7 *yuan* in 1993. Meanwhile the number of rural poor declined from 125 million to 80 million over the same period.

The success of the regional development strategy of poverty alleviation greatly encouraged the Chinese government to formulate further plans for lifting all the rural poor out of poverty. In 1994, the government proclaimed the implementation of '8-7' Poverty Reduction Plan, aiming to lift the remaining 80 million poor out of poverty in the coming seven years. In this plan, the government proposed a concrete objective that one labourer per household would be transferred to off-farm work. Major measures included the development of labour-intensive industries and township and village enterprises in poverty-stricken areas, the inter-regional transfer of rural labourers organised by government agencies, and the spatial relocation of poor households living in extremely poor and remote areas. Moreover, the development of education, training and technological extension was encouraged as a way to improve farmers' qualifications and their employability in non-agricultural sectors.

In 1994, the Chinese government adjusted the national poverty criteria for the state-designated poverty counties to include counties with annual per capita rural incomes of less than 400 *yuan*. According to this standard, 592 counties were officially designated as poverty counties, and their population accounted for 72 per cent of the total rural poor population in China. At the same time, most provincial governments also set up local criteria and designated a number of counties as official provincial level poverty counties. The wide social participation in anti-poverty actions yielded results. The number of rural poor in the state-designated poverty counties fell from 58.6 million in 1994 to 17.1 million and the rural poverty incidence fell to less than 3 per cent by the end of 2000, thereby achieving the goals of the '8-7' Poverty Reduction Plan (see Table 1.1).

At the beginning of the new millennium, the Chinese government announced a rural poverty alleviation programme for the period 2001 to 2010, which targeted the marginalised poor populations concentrated in remote and mountainous regions. In this ten-year programme, labour transfers and reallocation were also emphasised as an important means to implement the regional development and anti-poverty strategy. According to an interim assessment report released by the State Council Leading Group Office of Poverty Alleviation and Development in 2006, the implementation of the 2001–2010 Rural Poverty Alleviation Program has made great achievements. At present, 45,100 thousand villages have already been chosen for incorporation into an integrated poverty alleviation programme. A training network at the national, provincial, city and county levels has been constructed with nearly 800 training bases, which has trained 3.18 million rural labourers from poor households and helped them to find employment in non-agricultural sectors. Moreover, around 1.5 million rural poor who live in extremely vulnerable areas have been reallocated (State Council Leading Group Office of Poverty Alleviation and Development 2006).

Since 2003, two national programmes have been in operation and will continue until 2010. One provides training for around 60 million rural migrants to improve their employability in non-agricultural sectors; the other is to provide technical training for about 16 million farmers in new cultivation methods. These programmes are expected to greatly benefit rural poverty reduction and rural development.

Migration and the poverty trap

Characteristics of poor households

Income (consumption) poverty measures the status of poverty by comparing family income and consumption expenditures with a given poverty line.

This measurement is virtually an *ex post* methodology, because income (consumption) is the outcome of family economic activity. Poverty is multi-dimensional. Factors to directly or indirectly affect the process of family income generation include family assets, education, health status, local infrastructure, natural disasters, access to public services and participation in social activities. In most cases, income (consumption) poverty directly relates to assets-based and capability poverty that reflects the status of individual deprivation and social exclusion (Sen 1992; World Bank 2006). The assets-based and capability poverty is often the major cause of chronic income (consumption) poverty and a vicious poverty cycle.

Differences in income and consumption between poor and non-poor households mainly derive from their differences in assets and human capital. Table 1.6 provides evidence for this comparison. As shown in the table, per capita income of the poor and low-income households in 2004 is 578.7 *yuan* and 853.5 *yuan*, equivalent to only 19.7 per cent and 29.1 per cent of the national average per capita income of rural households respectively. Agricultural income is the main source of income for poor households, accounting for 68.4 per cent. In contrast, the share of wage income in the national average per capita income is 34.0 per cent, 14.1 percentage points higher than in poor households. The slow income growth of the poor and low-income rural households means that a large share of income is devoted to food expenditures. The Engle coefficients of the poor and low-income households, which measure the ratio of food expenditures to total consumption expenditures, is 71.3 per cent and 66.5 per cent, equivalent to 151.1 per cent and 140.9 per cent of the national average, respectively. Per capita living space and ratios of households with safe drinking water also indicate significant differences in the quality of life between poor and non-poor households (Table 1.6).

Three categories of factors cause income and consumption differences between the poor and non-poor households. First is the difference in agricultural production and market activities. Low-income levels and financial difficulties constrain the capability of the poor households to increase their input and productive investment. The semi-commercial characteristics of agricultural production mean that poor households produce agricultural goods for sale and personal consumption, which reduces their opportunities to make money by fully participating in market activities. In Table 1.6, per capita input of poor households is equivalent to less than half the national average, while the per capita productive fixed assets investments of poor households are equivalent to less than one-third of the national average; comparable figures for low-income households are slightly more than half and more than one-third of the national average, respectively. The commodity rates of staple agricultural products like grain, cotton and oil plants, vegetables and fruit from poor and low-income households are all below the national average, with the exception of the commodity rate of cotton from poor households.

Second is the difference in adult education. Adult illiteracy rates among poor and low-income households are respectively 1.5 and 1.3 times higher than the national average. Average school years in poor and low-income households are 6.4 and 7.0 years, equivalent to 81 and 89 per cent of the national average, respectively.

Table 1.6 Characteristics of the poor and low-income households

Item	2004			National = 100		Change rate between 2000 and 2004		
	National	Poor	Low income	Poor	Low income	National	Poor	Low income
Household size (person)	4.1	5.4	4.9	131.7	119.5			
Income, consumption and life quality								
Per capita income (yuan)	2936.4	578.7	853.5	19.7	29.1	30.3	12.3	5.7
Agr. income (%)	47.6	68.4	48.4	143.7	101.7	-1.7	3.8	-26.1
Wage (%)	34.0	19.9	22.7	58.5	66.8	9.0	-2.9	3.2
Per capita consumption expenditures (yuan)	2,185.0	602.0	822.0	27.6	37.6	30.8	10.6	8.5
Engel coefficient (%)	47.2	71.3	66.5	151.1	140.9	-3.9	3.5	2.9
Per capita housing value (yuan)	6,307.9	2,178.6	2,625.6	34.5	41.6	48.1	36.5	26.3
Per capita housing areas (sq. meter)	27.9	16.3	18.4	58.4	65.9	12.5	15.6	7.0
Ratio of safety drinking water (%)	69.9	55.3	58.0	79.1	83.0	8.0	7.0	11.1
Agricultural production and market activities								
Per capita production input (yuan)	923.9	384.2	463.7	41.6	50.2	41.2	21.8	20.2
Per capita purchasing fixed productive assets (yuan)	106.4	29.0	33.7	27.3	31.7	66.5	16.9	-14.7
Commodity rate of grain (%)	41.1	28.7	30.4	69.8	74.0	5.9	-10.0	-10.3
Commodity rate of cotton (%)	78.0	84.1	77.8	107.8	99.7	1.8	-4.2	-5.5
Commodity rate of oil seeds (%)	55.1	45.9	45.6	83.3	82.8	13.6	-4.8	-6.4
Commodity rate of vegetables (%)	59.2	29.8	34.0	50.3	57.4	34.2	0.3	21.4
Commodity rate of fruits (%)	78.8	55.0	64.5	69.8	81.9	14.4	-0.2	-2.0

Continued

Table 1.6 Cont'd

Item	2004			National = 100		Change rate between 2000 and 2004		
	National	Poor	Low income	Poor	Low income	National	Poor	Low income
Children and adult education								
Children enrollment rates aged 7–15 (%)	97.3	92.7	95.5	95.3	98.2	3.4	9.1	5.3
Adult illiteracy rates (%)	10.2	22.5	17.7	220.6	173.5	−11.3	−11.8	−5.9
Average schooling years of labor forces	7.9	6.4	7.0	81.0	88.6	2.6	4.9	2.9
Illiteracy and primary school of labour force (%)	36.7	56.7	48.7	154.5	132.7	−8.9	−6.9	−4.9
Middle school of labor forces (%)	50.4	37.6	43.3	74.6	85.9	4.8	11.9	3.3
High and specialized school of labor force (%)	12.1	5.3	7.9	43.8	65.3	9.0	−1.9	16.2
College and above of labor forces (%)	0.8	0.4	0.2	50.0	25.0	60.0	300.0	0.0
Community facilities								
Ratio of villages that have highways (%)	98.0	95.4	96.4	97.3	98.4	2.9	5.5	6.3
Ratio of villages that have telephones (%)	97.0	88.8	91.3	91.5	94.1	9.6	31.8	21.6
Ratio of villages that can watch TVs (%)	99.3	97.8	98.7	98.5	99.4	1.2	3.2	2.8
Ratio of villages that have elasticity (%)	99.4	96.7	98.5	97.3	99.1	0.7	2.2	2.1

Source: Rural Survey Department of the National Bureau of Statistics (2005).

The educational distribution of the labour force further illustrates the significant difference in education levels in poor and non-poor households. More than 60 per cent of labourers in poor households have not graduated from middle school, against the national average of 44.5 per cent. The disadvantage of less education in the poor households has prevented their members from improving agricultural productivity or from being employed in non-agricultural sectors.

The third factor is the differences in local infrastructure and access to public services. Rural infrastructure includes roads, irrigation, electricity, communication, transportation and so on, which play an important role in agricultural production and living conditions. As shown in Table 1.6, rural infrastructure and public utilities in poor areas are well below the national average. Even so, state investment in poor areas has helped to narrow the gap between the poor and non-poor areas.

Effectiveness of migration in reducing poverty

According to development theory, the poor caught in the poverty trap face a set of vicious cycles, which keep them mired in poverty. Investment in health and education is central to enable the poor to benefit from the interlocking set of self-reinforcing virtuous cycles and escape from the poverty trap.

Migration is part of a process of human capital investment. Only minimum financial input and psychological preparation are needed to cover the cost of job hunting and transportation, and to deal with the uncertainties and risks in the migration process. Studies show that the extremely poor are less likely to migrate due to their lack of money, information and education, and are more risk-averse, but they will pursue migration if given help in training and finding jobs. The monitoring figures show an upward trend in migration from poor areas. In 2001, 11.8 per cent of total labourers in officially designated poor counties left to work outside and this number had risen to 16.6 per cent by 2004. The increasing inter-provincial migration from the officially designated poverty counties indicates an improvement in the migration capability of rural labourers in poor areas (see Table 1.7).

Remittances are one of the most important means by which migration contributes to poverty reduction. For example, many surplus rural labourers in the western provinces have found jobs in the more developed areas of their own provinces or in the coastal provinces, with the assistance of active programmes. Many send home remittances that allow relatives on the farms to improve their living standards, or else they bring money back home to set up small businesses, creating needed jobs in the villages. It is reported that remittances from outside migrants to Sichuan province amounted to an estimated 20 billion *yuan* (US$ 2.4 billion) in 1995, accounting for 7 per cent of the province's GDP. Around 30,000 peasants who returned to that province have started their own businesses, creating thousands of local jobs. Remittances from migration were 2.8 times the per capita income of the poor households in the officially designated poverty counties (Table 1.7).

Using rural household survey data for four poverty counties, Cai and Du (2005) analysed the impact of remittances on poverty reduction. According to

Table 1.7 Migration in the state designated counties

	2004	*2001*
Mobility channels		
Gov't and organization	3.4	3.82
Relatives and friends	38.9	41.93
Self own	57.7	54.25
Region distribution		
Outside township and within county	12.6	22.74
Outside county and within province	21.1	24.47
Inter-province	66.3	52.79
Outside duration		
0–6 months	36.8	na
6 months and over	63.2	na
Income		
Total income	3921.6	3268.53
Self-own consumption expenditures	1447.4	1344.85
Remittances	1611.8	1706.11
Number of migrants		
Share of rural labor forces	16.6	11.8

Source: Rural Survey Department of the National Bureau of Statistics (2002, 2005).

the data, the per capita income of migrants was 2,907 *yuan* before remittances, while that of other family members was 602 *yuan*. On average, migrants remitted one-third of their income per person (980 *yuan*), with other family members receiving 465 *yuan* per person, or 77 per cent of their own per capita earnings. Using the rural poverty line of 635 *yuan* in 2000, they found that the poverty incidence of migrants was 17.5 per cent before remittances, and 27.8 per cent after remittances, whereas the poverty incidence of other family members was 67.1 per cent before and 49.2 per cent after receiving remittances. Such evidence suggests that migration helps poor families to deal with and eventually escape poverty.

Ways to eradicate chronic poverty

As the left-behind rural poor are increasingly marginalised, an analysis of their income dynamics provides useful information for understanding poverty persistence. However, this method has a strict requirement that the data should be a repeated household panel so that the index of aggregate poverty can be decomposed into chronic and transitory poverty. Rodgers and Rodgers (1993) used the Panel Study of Income Dynamics data, and found that during the 1970s and the mid-1980s, chronic poverty in the US was a more serious problem than transitory poverty. Based on a 1990–1995 panel dataset, Jyotsna and Ravallion (1998) found that consumption variability accounted for a large share of observed poverty in rural China. They suggested that China's anti-poverty policies should place greater emphasis on the problem of transitory poverty.

The Rural Survey Department of the National Bureau of Statistics (2002) re-examined the issue of rural poverty persistence by using large-sample panel data from 1997 to 2001 in the officially designated poverty counties. The results obtained from 16,000 continuous households[3] confirmed that transitory poverty is the dominant form of poverty and accounts for 91.7 per cent (65.1 per cent) of rural income (consumption) poverty if using the official poverty line (see Table 1.8). However, if the low-income poverty line (equivalent to one-dollar-per-day criteria) is used, the share of income transitory poverty dropped to 79.9 per cent, whereas the share of consumption transitory poverty fell to 44.8 per cent, which indicates that we should be cautious when examining the nature of rural poverty.

Family size, education and geographic location have a significant impact on headcount index and poverty persistence. As shown in Table 1.8, the headcount index rises with the increase in family size, but decreases with the increase of per capita education levels. The headcount index in the eastern region is around one-quarter of that in central and western regions. Like the headcount index, the chronic poverty index and its share in the headcount index have a similar relationship with family size, per capita education and geographic location.

The factors that cause transitory and chronic poverty differ. Families caught in transitory poverty often experience the temporary shocks of natural disaster, agricultural price fluctuations, variations in the economic business cycle, short-time illness, temporary unemployment, and so on. They can recover from those temporary shocks, and their income and consumption recover over time. In contrast, families caught in chronic poverty often face long-lasting adverse factors such as living in remote areas with bad natural environments, poor production conditions and low agricultural productivity, family members with serious illness, and low participation in market activities and non-agricultural employment. It is difficult to improve the quality of life of people in these households through agricultural production, and it is difficult for them to recover from external adverse shocks.

Different sources of transitory and chronic poverty have different implications for policy and intervention. If transitory poverty is the major component of poverty, a well-funded social safety net would be a good tool to help the poor overcome temporary shocks. If poverty is composed largely of chronic poverty, the developmental approach such as strengthening asset accumulation, human capital investment, infrastructural investment and the provision of public services will be an effective and cost-efficient option to eliminate poverty in the long run. For some poor populations who live in isolated remote border or mountainous areas, government-sponsored relocation would be a sound choice to cut down the overwhelming cost.

Cai and Du (2005) used endowments as an instrument to predict the likelihood of households in poor areas participating in migration. They then grouped these households into three types: households with high endowments which tend to be more responsive to migration; households with low endowments which tend to be less responsive to migration; and households at the cut-off point which have moderate endowments and tend to be sensitive to migration because the

Table 1.8 Transitory and chronic poverty in the state designated counties: 1997–2001

	Official poverty line: 630 yuan				Low income line: 830 yuan			
	Headcount index	Chronic poverty	Transitory poverty	Share of transitory poverty	Headcount index	Chronic poverty	Transitory poverty	Share of transitory poverty
Income poverty	0.0770	0.0064	0.0707	91.70	0.0745	0.0150	0.0596	79.90
Household size								
1	0.0111	0.0041	0.0071	63.50	0.0299	0.0099	0.0200	66.92
2	0.0128	0.0033	0.0095	74.54	0.0346	0.0147	0.0199	57.59
3	0.0117	0.0028	0.0089	75.79	0.0338	0.0149	0.0189	55.98
4	0.0139	0.0040	0.0099	71.33	0.0421	0.0207	0.0214	50.90
5	0.0214	0.0078	0.0136	63.35	0.0586	0.0341	0.0245	41.84
6	0.0289	0.0160	0.0173	59.98	0.0769	0.0475	0.0295	38.30
7	0.0357	0.0185	0.0172	48.08	0.0910	0.0642	0.0268	29.33
8	0.0413	0.0180	0.0233	56.49	0.0122	0.0691	0.0331	32.34
Education								
Illiteracy	0.0361	0.0166	0.0196	54.09	0.0902	0.0594	0.0308	34.15
Semi illiteracy	0.0228	0.0079	0.0149	65.26	0.0629	0.0359	0.0270	42.93
Primary school	0.0181	0.0058	0.0123	68.01	0.0499	0.0262	0.0237	47.52
Middle school	0.0157	0.0051	0.0106	67.36	0.0453	0.0237	0.0215	47.56
High school	0.0132	0.0043	0.0089	67.72	0.0392	0.0199	0.0193	49.20
Specialized high school	0.0082	0.0020	0.0062	75.17	0.0250	0.0099	0.0151	60.40
College and above	0.0072	0.0010	0.0061	85.60	0.0275	0.0115	0.0159	58.04
Region								
East	0.0055	0.0008	0.0047	85.66	0.0195	0.0068	0.0128	65.37
Central	0.0185	0.0069	0.0116	62.67	0.0513	0.0285	0.0228	44.39
West	0.0199	0.0067	0.0132	66.44	0.0555	0.0309	0.0245	44.21
Consumption poverty	0.0184	0.0067	0.0120	65.1	0.0515	0.0284	0.0230	44.8

Source: Rural Survey Department of the National Bureau of Statistics (2002).

predicted income at the cut-off point is almost equivalent to the official poverty line. From the perspective of migration there is a set of specific but different policy measures for government intervention. For the first household type, a better institutional environment is very important for them to improve their quality of life through free mobility. For the second type of households, the provision of basic needs is necessary for them to escape poverty. Strengthening household assets accumulation, human capital investment and the provision of public services can increase their endowments, enhance their capacity to migrate and enable them to cast off poverty in the long run. As for the third type, providing training, employment information and services are mostly helpful in enabling them to grasp migration opportunities.

Migration and the urbanisation of poverty

Determinants of the urbanisation of poverty

The urbanisation of poverty is largely determined by the rate of rural–urban migration and the growth of employment in urban sectors. If the growth of employment outpaces the rate of rural–urban migration, it is possible to achieve the dual goal of both poverty alleviation and urban development. If the process of migration is hindered by institutional and policy barriers, the segregation of the urban labour market will distort the efficient allocation of resources and factors of production. Labour market discrimination and social exclusion will force most rural migrants to work in the informal sectors and to choose informal settlement in order to reduce their living costs in the urban areas, a tendency which will probably exacerbate the problem of urban poverty in the future.

According to the report of the United Nations Human Settlements Programme (2003), the world's urban population increased by 36 per cent during the 1990s. If such growth rates are sustained in the future, poverty will become increasingly urbanised across the globe due to the lagged development of urban infrastructure, job creation and public services. At present, at least 1 billion people worldwide live in urban slums. Moreover, in 30 years time, one in every three will live in urban slums characterised by poor public health, lack of basic infrastructure, inadequate public services and widespread violence and insecurity.

Ravallion (2001) verified that owing to the rapid transfer of poverty from rural to urban areas via migration, the growth rate of urban pauperisation in developing countries outpaces the speed of urbanisation itself. According to data from 39 developing countries, he found that the rate of urban pauperisation is 26 percentage points higher than the rate of urbanisation. If this momentum persists and global urbanisation reaches 52 per cent in 2020, the proportion of the urban poor as a percentage of the total urban population will rise to 40 per cent.

The increase in the proportion of the poor in urban areas will exacerbate the cost of development. Urban slums are often excluded from urban planning programmes, receive little productive public investment, and suffer from a lack of income-generating opportunities. Residents in informal settlements also face a high degree of uncertainty, since their rights to remain in their homes are often poorly defined. As a result, migrants are particularly prone to become victims of urban pauperisation. To avoid this outcome would require that greater policy attention be directed towards the living and employment conditions of rural migrants.

Re-estimating urban poverty

Several studies show that urban poverty will increase if migrants are included in calculations as migrants are often excluded from affordable public housing, health services and schooling for children. However, results from a new China Urban Labour Survey (CULS) jointly conducted by the Institute of Population and Labour Economics of the Chinese Academy of Social Sciences and the World Bank in 2004 and 2005, do not fully support this conclusion.

As shown in Table 1.9, we chose different poverty criteria to measure the poverty of urban residents and rural migrants. The diagnostic *dibao* (minimum living standard scheme) poverty line is 1,982 *yuan*, the low-income poverty line is 1,112 *yuan*, the one-dollar-per-day poverty line is 1,124 *yuan*, and the two-dollars-per-day poverty line is 2,247 *yuan*. Thus, the estimates between the diagnostic *dibao* poverty line and the two-dollars-per-day poverty line, and between the low-income poverty line and the one-dollar-poverty line are close. The estimates of migrants' poverty incidence from low-income poverty and one-dollar-per-day poverty lines are close to 1.4 percentage point above those of urban residents, but figures from the *dibao* poverty line and the two-dollars-a-day poverty line for urban residents and rural migrants are almost equal. If we include rural migrants

Table 1.9 Re-estimation of urban poverty

City	Poverty line			
	Dibao line	Low income line	One-dollar-a-day	Two-dollars-a-day
Poverty incidence of urban residents				
5 large cities	3.5	1.4	1.4	4.1
5 small cities	7.0	3.1	3.1	8.4
Total	5.3	2.2	2.3	6.3
Poverty incidence of rural migrants				
5 large cities	3.7	2.3	2.3	4.0
5 small cities	7.1	3.9	3.9	8.2
Total	5.4	3.1	3.1	6.2
Poverty incidence of urban residents and rural migrants				
5 large cities	3.5	1.8	1.8	4.0
5 small cities	7.1	3.4	3.4	8.3
Total	5.4	2.6	2.6	6.3

Source: Institute of Population and Labour Economics, CASS, China Urban Labour Survey, 2005. This survey includes five large cities (Shanghai, Wuhan, Shengyang, Fujian, and Xian), and five small surrounding cities (Wuxi, Yichan, Benxi, Zhuhai and Baoji). In each large city, about 500 urban and migrant households are surveyed. In each small city, about 400 urban and migrant households are surveyed.

in the measurement of urban poverty, the urban poverty incidence is only slightly increased by a 0.1 percentage point.

Higher human capital, higher mobility and lower unemployment rates explain the relatively low poverty of migrants in cities. Migration is a process of natural selection. Under the restrictions of the *hukou* system, only migrants with better human capital are able to penetrate the urban 'invisible wall'. A number of studies show that rural migrant workers are primarily young individuals with, on average, one school year more than those who do not choose to migrate, and equivalent to that of urban residents. If migrants cannot find a job in one city, they can try their luck in another one. If they fail in urban areas, they can return to farming. Their higher mobility also ensures them a low unemployment rate. According to the statistics of the fifth population census in 2000, the unemployment rate of migrants was 3.6 per cent, compared to 9.1 per cent for urban residents. The China Urban Labour Survey also confirms the 2005 unemployment rate among migrants at 2.7 per cent, against 8.6 per cent for urban residents.

Labour market discrimination and social exclusion

Although as yet rural–urban migration in China has not had much of a negative impact on urban poverty, the *hukou* system remains a fundamental barrier to migration due to the incomplete nature of the reform (Roberts 2000). Comparing the Chinese urban restrictions targeting rural migrants with the stringent policy measures adopted by Germany and Japan to limit immigration, Solinger (1999) finds that in terms of entry rules, citizenship rights and treatment, the former are more restrictive than the latter.

Evidence illustrates that rural migrants lack the necessary social protection and have low social security coverage and limited access to urban public services. Many rural migrants work in harsh conditions all the year round only to find that they cannot get paid. For work units employing migrant workers in 2001, the default ratio was 12.02 per cent, considerably above the 8.59 per cent for those employing only urban residents (Cai and Wang 2005). According to the China Urban Labour Survey, in all work units with migrant workers, less than 10 per cent of migrant workers are provided with old-age social security, while more than 70 per cent of their urban counterparts enjoy this security; less than 10 per cent of migrant workers are covered by medical insurance against more than 65 per cent of urban resident workers. Migrant workers have hardly any chance of receiving formal education after entering the city, and they have to pay higher tuition fees for their children's education. The difference in tuition for students with and without local *hukou* was around 30 per cent in 2005.

Conclusions

China has achieved remarkable progress in poverty alleviation since the start of the reforms in 1978 by initiating regional development strategies and encouraging broad social participation in economic growth. Rapid agricultural growth in the initial reform stage reduced the incidence of rural poor by half, but since the mid-1980s it slowed down, leading to the deceleration of rural poverty reduction and widening income inequality.

With the growing regional concentration of rural poverty, migration has played an increasingly important role in rural income growth and poverty reduction. Much experience has been gained by Chinese planners and government officials in ways of promoting economic development by abolishing institutional barriers and correcting economic structural distortions. This has enabled migration to become more responsive to regional income inequalities and to accelerate the transformation of economic structures. In particular migration has strengthened the linkage between rural and urban areas through the contribution of remittances to the rural poor: in 2004 remittances accounted for 18 per cent of rural income and reduced rural poverty by nearly 20 percentage points. Migration has also strengthened rural–urban linkages by providing cheap rural labour for urban economic growth and by narrowing rural–urban disparities through the reallocation of factors of production. Yet even though migration is important for poverty alleviation and economic development, it nevertheless remains difficult for extremely poor households with low endowments to take advantage of labour markets and to benefit from rapid economic growth. The declining employment elasticity is related to the slowdown in the rate of rural poverty reduction and the increase in urban poverty. This suggests that the coordination of economic growth and employment is important to reduce both rural and urban poverty.

At present, massive migration does not significantly worsen the incidence of urban poverty, but labour market discrimination and social exclusion have increased the risks and vulnerabilities of rural migrants. In order to avoid the urbanisation of poverty, abolishing various remnants of the *hukou* system and employment policy constraints on labour mobility, as well as establishing a portable social security system for migrants, are key to enabling them to grasp the opportunities from rapid economic growth without falling into poverty.

The increasing marginalisation of rural poverty has alerted policy-makers to place greater emphasis on the establishment of a rural social security system. The minimum standard of living scheme (*dibao*) and the new rural cooperative medical scheme should be priorities since so many marginalised people are extremely poor and illness is a key cause of their poverty. A pension system is almost non-existent in rural China. The gradual creation of a pension system in rural areas is extremely important to reduce the poverty incidence of the rural elderly. Moreover, the social development approach of strengthening the capacity for asset accumulation, human capital investment as well as the provision of public services should be pursued to enhance the capability of extremely poor households and lift them out of chronic poverty.

Notes

1 As the most populous developing country in the world, arable land in China is a mere 0.1 hectare per capita, only half the world's average.

2 All agricultural products were classified into three categories. The first category of agricultural products includes grains, cotton and oil seeds; the second category of agricultural products includes pork, beef, mutton and fishes; the rest belongs to the third category of agricultural products. The state asked farmers first to meet the requirement of quota delivery for the first and second categories of agricultural products, and then farmers could sale their residual at local retail markets.

3 In each year, the total sample size of poverty monitoring is 50,000 households, but only 16,000 households are repeated in the continuous five years.

References

Asian Development Bank (ADB) (2004) 'The suggestions to set up *Dibao* in rural area', Internal report, Asian Development Bank.

Cai, F. and Wang, D. (2003) 'Migration as marketization: what can we learn from China's 2000 census data?' *The China Review*, 3(2): 73–93.

Cai, F. and Wang, D. (2005) 'Impacts of domestic migration on economic growth and urban development in China', paper presented at the Migration and Development Within and Across Borders: Concepts, Methods and Policy Considerations in International and Internal Migration Conference, International Organization for Migration (IOM), New York, November 17–19.

Cai, F. and Du, Y. (2005) 'Changing nature of rural poverty and new policy orientations', working paper for the Institute of Population and Labour Economics, Chinese Academy of Social Sciences.

Chen, S. and Wang, Y. (2001) 'China's growth and poverty reduction; recent trends between 1990 and 1999', working paper, Policy Research Department, World Bank, Washington DC.

Chen, X. and Zhang, H. (eds) (2005) *Building an Equal Employment System for Rural Labourers*. Beijing: China Financial and Economic Press.

China's News Office of the State Council (2001) 'China's Rural Poverty Reduction and Development'. Available at: http://www.china.org.cn/ch-book/fupinkafa/f1.htm

Huang, J., Rozelle, S. and Qi, Z. (2005) 'Macroeconomic policies, trade liberalization and poverty in China', in Guoliang, W. (ed.) *A Policy Study on the Poverty Reduction Program of PRC: Trends and Challenges*. Beijing: Social Sciences Academic Press.

Hussain, A. (2003) 'Urban poverty in China: measurement, patterns and policies', International Labour Office, Geneva.

Jyotsna, J. and Ravallion, M. (1998) 'Transient poverty in post-reform rural China', *Journal of Comparative Economics*, 26(2): 338–357.

Khan, A.R. (2000) 'A comparative analysis of selected Asian countries', United Nations Development Program. Available at http://www.undp.org/poverty/publications/case/macro/asia.doc

Khan, A.R. (1998) 'Poverty in China in the period of globalization: New evidence on trends and patterns', Issues in Development Discussion Paper No. 22, Development Policies Department, International Labour Office, Geneva.

Li, S. (2001) 'The increasing urban poverty at the end of the 1990s and its reason', Institute of Population and Labour Economics, CASS, China Urban Labor Survey, 2005. Available at: http://www.cass.net.cn/jingjisuo/yjlw/01.asp?id=194

Li, S. and Yue, X (2004) 'A survey on China's rural urban disparity', *Finance*, 4: 30–38.

Lin, J., Gewei, W. and Yaohui, Z. (2004) 'Regional inequality and labour transfers in China', *Economic Development and Cultural Change*, 52(3): 587–603.

Lin, J. and Li, Y. (2005) 'China's poverty reduction policy: trends and challenges', in Guoliang, W. (ed.) *A Policy Study on the Poverty Reduction Program of PRC: Trends and Challenges*. Beijing: Social Sciences Academic Press.

Ministry of Agriculture (MOA) (2004) *China Township Enterprises Yearbook*. Beijing: China Agricultural Press.

National Bureau of Statistics (NBS) (2005) *China Statistical Yearbook*. Beijing: China Statistics Press.

National Bureau of Statistics (NBS) (2006) *China Yearbook of Rural Household Survey*. Beijing: China Statistics Press.

Rural Survey Department of National Bureau of Statistics (2002) 'Poverty monitoring report of rural China', China Statistics Press, Beijing.

Rural Survey Department of National Bureau of Statistics (2005) 'Poverty monitoring report of rural China', China Statistics Press, Beijing.

Ravallion, M. (2001) 'On the urbanization of poverty', working paper, Development Research Group, World Bank, Washington DC.

Ravallion, M. and Chen, S. (2004) 'China's (uneven) progress against poverty', Policy Research Paper 3408, Development Research Group, World Bank, Washington, DC.

Roberts, K. (2000) 'Chinese labour migration: Insights from Mexican undocumented migration to the United States', in West, L. and Yaohui, Z. (eds) *Rural Labour Flows in China*. Berkeley: Institute of East Asian Studies.

Rodgers, R.J. and Rodgers, J.L. (1993) 'Chronic poverty in the United States', *Journal of Human Resources*, 28(1): 25–54.

Sen, A. (1992) *Inequality Re-examined*. Oxford: Clarendon Press.

Solinger, D.J. (1999) Citizenship issues in China's internal migration: Comparisons with Germany and Japan. *Political Science Quarterly*, 114(3): 455–478.

State Council Leading Group Office of Poverty Alleviation and Development (2006) 'The interim policy assessment report of the 2001–2010 China's rural poverty alleviation and development', State Council, Beijing.

United Nations Human Settlements Programme (2003) *The Challenge of Slums: Global Report on Human Settlements*. London: Earthscan Publications and UN-Habitat.

Wang D., Fang, C. and Wenshu, G. (2005) 'Globalization and internal migration in China: New trends and policy implications', paper presented at Monash University, Australia.

World Bank (1992) 'China: Strategies for reducing poverty in the 1990s', World Bank, Washington DC.

World Bank (2001) 'China: Overcoming rural poverty', World Bank, Washington DC.

World Bank (2006) 'World Development Report 2006: Equity and Development', World Bank, Washington DC.

2 Migrant remittances in China

The distribution of economic benefits and social costs

Rachel Murphy

The term 'remittances' refers to the money that migrant workers send back to their communities of origin. Remittances are an integral feature of the migration system in China. Remittances occur largely because migration forms part of a strategy for 'rural livelihood diversification'. This means that rural households spread their earning activities over a range of on-farm and off-farm activities in order to minimise their risks and raise their returns to available labour. The world over, off-farm activities generate more income than agriculture and it is access to this cash rather than the size of land allocations that determines wealth inequalities within rural communities. In localities where the local economy is unable to provide off-farm employment, it becomes necessary for household members to find work in cities (Hare 1999).

In urban labour markets, rural people are generally relegated to dirty, dangerous and poorly paid occupations that offer few prospects for advancement to more comfortable and stable living and working conditions. Owing to their precarious and poor living and working conditions, most migrants leave family members, especially their children, in the villages. Indeed in 2000 only 7 per cent of rural–urban migration was family migration (National Bureau of Statistics 2002, cited in Tao and Xu 2007). In such circumstances, migrants commonly perceive that their stay in the city is not permanent. Many migrants therefore remit money not only to care for their rural family members but also to maintain a stake in the rural community and its resources for when illness, the lack of an urban job, family circumstances or old age force them back to the village (Bai and He 2002). Studies report average durations of absence ranging from four to seven years, with some evidence that family obligations cause women to return sooner (Ma 2001; Murphy 2002; Ngai 2005).

According to a report released by the Consultative Group to Assist the Poor,[1] China's rural migrants sent home nearly US$ 30 billion in 2005 (Cheng and Zhong 2005: 4). To provide context, this sum is more than the amount that China or any other country receives from international cross-border flows. For instance, India, the world's largest recipient of money transfers from overseas migrant workers is estimated to have received US$ 22 billion in 2005, with China close behind (de la Torre 2005). The significance of domestic remittances is also apparent when one considers the huge numbers of people that receive them. Owing to the shorter distances of travel, the cheaper costs of labour market entry and the larger volume of domestic migrants as compared with international migrants, domestic remittances are likely to give benefits to greater numbers of poor people than international money transfers. Clearly, and as is also reported by Wang and Cai in this volume, in the case of China, remittances have greatly improved the incomes of rural people. One 2004 source finds that remittances contributed 18 per cent of total rural income (Huang and Zhan 2005b). Meanwhile, a study conducted in 2005 found that domestic remittances contributed between 20 and 50 per cent of the total income of recipient rural households (Cheng and Zhong 2005: 3).

In order to understand the contributions of remittances to social and economic development, this chapter explores several questions: How is this money distributed across and within regions? What channels are used to send the money back to the rural areas? How are remittances spent? Who in the rural community gets the money? Why do some migrants fail to remit? What are the policy implications of how the money is remitted, distributed and used?

In the process of answering each question, this chapter confirms an orthodox view in mainstream migration policy circles that remittances improve rural livelihoods by enabling the people left behind to consume items which improve their quality of life. In the case of China this benefit of remittances is enhanced by the widespread participation of people from poor regions in migration, by the accessibility of urban labour markets, and by the widespread availability of reliable channels for remitting money. Yet this chapter also calls for attention to a reality often overlooked by planners and migration scholars – remittances are obtained at considerable social cost to migrants and their families, and both the economic benefits and social costs of migration are distributed unequally.

Regional distribution of remittances

A range of movements across and within provincial boundaries means that many localities within China benefit from remittances. According to the *China Daily* (15 May 2004) intra-provincial migration accounts for 50 per cent of labour migration. Most intra-provincial migrants move from rural to urban areas in the eastern provinces, where they can earn sizeable amounts of money to send back to their hometowns. Farmers in poorer interior provinces also migrate to nearby cities. But these destinations offer fewer earning opportunities and lower incomes than coastal destinations, so attract less than 30 per cent of migrants from the interior provinces. The majority of farmers from the densely populated interior provinces prefer instead to migrate to the eastern provinces (Cai 2005). Not surprisingly, therefore, these interior provinces receive the greatest volume of remittances. In 2000, Sichuan province's remittances reached 202 billion *yuan*, equivalent to its total fiscal revenue. The same year, Anhui province's remittances reached 174.3 billion *yuan*, exceeding the total amount of its fiscal revenue by 43 billion *yuan* (Wang 2003: 3).

Although migration has generated a substantial return flow of money to rural areas, the remittances have not been sufficient to counter China's rural–urban gap, which has been growing and is now one of the largest in the world. If subsidies are included, rural people earn one-third the income of urban residents in general, and one-fifth the income of urban residents on the east coast. It may be the case, though, that these gaps would have been even wider without remittances (Huang and Zhan 2005b).

While remittances have not turned the tide of rural–urban inequality in China, there is nevertheless persuasive evidence to suggest that income from temporary migration has exerted an equalizing effect *within* rural areas. One study based on a panel survey of over 7,000 rural households conducted between 1987 and 1999 finds extensive and persistent inequality within the same rural regions, provinces and villages, and attributes this to unequal access to a growing local off-farm sector. Meanwhile, over the same time period, the authors observe that the earnings generated through temporary labour migration to locations outside the households' home counties have been relatively equalizing (Benjamin *et al.* 2005).

A study by Riskin – also based on panel data – reports a significant reduction in income inequality within rural areas during the years 1995 to 2002 (Riskin 2007). Riskin's findings are compatible with those of de Brauw *et al.* (2002) and Rozelle *et al.* (1999a) in that he attributes this reduction to the rapid expansion in rural people's access to wage employment that has occurred because of the emergence of rural labour markets, the growth in the rural private sector and a substantial increase in labour migration. Riskin points out that rapidly increasing labour mobility is demonstrated by the fact that in 1995 only 3 per cent of rural wage earners worked outside their home province, whereas by 2002 some 18 per cent did. Indeed, by 2002 as many as two-thirds of all rural wage earners were working outside their home villages and one-third of these were working outside their home counties (Riskin 2007).

Even though the differing degrees of availability of local off-farm employment during the periods considered in the above-mentioned studies have led to differing verdicts on whether incomes within rural areas have become more or less equal, the authors nevertheless concur that, at least in aggregate, remittances exert an equalizing influence within rural China. At the bare minimum there is consensus among scholars that rural migration and remittances do 'not cause a deterioration in income distribution and might improve it' (Li 1999: 113). Moreover, within the poorest regions the impact of remittances on poverty alleviation is profound. This becomes evident when we consider that a migrant remits approximately 3,000 *yuan* a year, while in 2003 60 million households were under a poverty line of 825 *yuan* (Huang and Zhan 2005b). The likely equalising effect of migrant wages within rural China and their role in lifting households out of poverty supports the UNDP assertion that 'in effectively assisting wealth distribution, remittances help in achieving MDGs' such as the creation and sustainability of livelihoods (United Nations Development Programme 2005).

That said, the story becomes more complicated when we consider two studies which find that remittances exert a potentially disequalising impact at regional and community levels. The first, based on a 1995 survey of rural household income suggests that although remittances are equalising in richer provinces such as Guangdong, where it is the poorest households that send migrants, they appear to be disequalising in poorer provinces such as Sichuan where it is the middle income households that send migrants: the poorest do not have the income for the initial costs of migration while the richest can often earn more income at home (Li 1999). The second study, based on a 1999 survey of 451 rural households with out-migrants in Sichuan, and a 2000 survey of 493 migrants in Beijing, finds that even though the poorest migrants from the poorest regions and the poorest households remit a higher proportion of their incomes on a more regular basis than their more privileged counterparts, income inequalities in the poorer rural areas are not reduced. This is because remittances increase the income gap between those rural households with migrants and those rural households without migrants (Li 2001).

The exclusion of some poorer households from the benefits of remittances would suggest a need to ensure that migration or other off-farm income sources are an option for the poorest households in middling and poorer regions. Rozelle *et al.* found that the information linkages of chain migration (thereby reducing initial job search costs) and village level access to credit are two important factors in facilitating out-migration (Rozelle *et al.* 1999a: 389). Several successful pilot projects, particularly in poor western regions, have tried to overcome such obstacles to migration by providing a combination of job training, job introductions, financial literacy training and remittance services to intending migrants and their families (Wang and Cai, this volume; Zhan 2004). Yet, while these projects are laudable, they remain relatively isolated and lack the resources needed for scaling up. Furthermore, financial institutions are reluctant to give credit to rural people, precisely because they are seen as highly mobile and so unreliable (Fu 2005). Moreover, even though credit and information services may make migration and therefore remittances available to more poor rural people, there are also some households, which owing to reasons of demographic cycle, ill health or other obstacles, are nevertheless unable to send a migrant. In the absence of social security or income subsidies, the members of these households have few possibilities for escaping hardship.

How is the money remitted?

Given the role of remittances in family maintenance, it is important that these funds are sent safely and cheaply. Safe channels for remitting money are available in much of China and the high volume of money remitted through formal channels reflects this. Based on a survey of 400 migrant workers, a study by the Consultative Group to Assist the Poor finds that around three-quarters of China's domestic remittances are sent through China Post, commercial banks and rural credit cooperatives (Cheng and Zhong 2005: 5). The percentage breakdown of formal routes for sending remittances is as follows: the post office, 62 per cent; commercial banks, 32 per cent; and rural credit cooperatives, 5.5 per cent (Cheng and Zhong 2005: 6). China Post is able to dominate the remittance services market because it has an expansive computerized network covering many rural communities.

Despite the wide availability of remittance service providers, however, people who live in poorer localities that are situated away from county seats, larger townships and roads have experienced increased difficulties in obtaining access to remittance service providers. In recent years, there has been a rapid decline in the number of branch offices and distribution points of commercial banks in rural areas (Cheng and Zhong 2005: 6–7; Fu 2005). Also, since 1999/2000, even the Post Office, which provides the widest coverage of remittance service outlets in poor regions, has been forced to close some branches. This is because the 1999 restructuring in the communications sector freed China Telecom from its longstanding obligation to cross-subsidise postal operations, and the compensation from government finance departments has not been sufficient to compensate for the loss. The situation in Wanzai, a rural county located in the agricultural interior province of Jiangxi, is fairly typical. Since 2000, three out of a total of its thirteen post-office branches have been forced to close, and all of these were located in remote townships (interview, China Post manager, Wanzai, 5 July 2006).

In many places where remittance services are available the charges are not high. Even so, the costs tend to fall disproportionately on the shoulders of poorer people. The fee is usually 1 to 1.5 per cent of the total amount remitted. For workers from poor regions, this is around 30 to 50 *yuan*, equivalent to their monthly food allowance (Cheng and Zhong 2005). Further costs may be incurred in cases where post office staff impose higher fees, levy money at both the sending and receiving end of the remittance process (Fu 2005), or fail to disclose information about cheaper remittance options (Cheng and Zhong 2005). Another disadvantage faced by certain rural customers is that some post office branches enforce a compulsory savings policy on them. The poor, old and less well educated may be more vulnerable to these infringements because they may not be aware of their rights and may feel intimidated by formal settings, forms and procedures. Sending money through the card-based systems (prepaid, debit, credit card) of the commercial banks, such as the Agricultural Development Bank, the People's Construction Bank of China and the Industrial and Commercial Bank of China, offers migrants the cheapest and fastest method of remitting. Again, however, the poor and less well educated tend to be disadvantaged because they are less likely to know about or to use these services.

A paucity of financial services near to rural hometowns, a reluctance to pay remittance service fees, and a lack of familiarity with remittance services are all factors that inform some migrants' decision to remit their money through informal channels. Informal remittance channels usually involve the carrying of the money by migrants themselves, or else entrusting a friend or relative to carry the money on their return to the village. At this point, it is worth noting that a Global Call to Action Against Poverty (GCAP) survey's finding that only 25 per cent of money is remitted informally is lower than expected given that elsewhere in the world the amounts remitted through informal channels are at least as high as those sent through formal ones. Yet, as is commonly noted in the migration studies literature, informal remittances are notoriously difficult to calculate. This has caused some migration experts to speculate that the total volume of domestic remittances in China may be much higher than the above-mentioned GCAP estimate of US$ 30 billion (Jennifer Isern of GCAP, cited in de la Torre 2005).

The drawback of informal methods of remitting is of course that people can be vulnerable to theft. In order to enhance the security of migrants' money, it has been common for long distance bus companies in China to employ security guards, particularly at peak times such as Spring Festival. Even so, the migrants are still vulnerable when they wait at bus and train stations and when they travel from the stations to their places of residence. Rural people can also be vulnerable when they carry large sums of money over long distances between the branch of the nearest post office or credit cooperative and their home village. In noting the devastating effects of the theft of remittances on migrants and their families as well as the general absence of networked rural credit cooperatives in poorer regions, one Chinese commentator recommends that the Post Office should create a special safe and fast remittance channel for migrant workers over the Spring Festival period (*Hainan Ribao*, 12 January 2006). The possibilities for such innovations are suggested by the concerted actions of many post office branches in coordinating mobile remitting services during the period of the 2003 SARS quarantine (Wang 2003).

Over the past decade, financial institutions have responded to the remittance difficulties faced by some rural people by initiating projects designed to improve services. As early as 1997, branches of the Agricultural Bank located in particular townships made arrangements with selected branches of banks in districts of the destination cities in which migrants from the respective townships were concentrated. These arrangements enabled the 24-hour electronic transfer of cash. At the time, having a shiny new bank card was a matter of pride for spouses and elderly people remaining in the rural areas (He 1997). Since then ever more banks, including the Bank of China, the Industrial and Commercial Bank of China and the Construction Bank have established and diversified remittance-related services, often tailoring the content and marketing of existing bank-card products to appeal more directly to migrant workers and their families. Some of these products have been launched with great fanfare and media publicity, with relevant institutions claiming that they are finding ways to support the new leadership's emphasis on the ' three rurals' (*sannong*) policy of improving rural livelihoods. For instance, in 2005 the Bank of China heavily promoted a new bank card for migrant workers with hometowns in Guizhou, one of China's poorest provinces (Chen 2006; Gu 2006). There has also been a drive championed by the Bank of China to enhance the capacity of rural credit cooperatives to handle cross-regional financial transactions of small and medium amounts by incorporating eligible rural credit cooperatives into the national system for dealing with large transfers (Chen 2006).

While the representatives of financial institutions use the language of the 'three rurals' to explain their new remittance services, a less explicitly acknowledged motivation for their innovations is undoubtedly profit. China Post certainly sees the banks' initiatives as a source of competition and has responded by devising ways to improve its own remittance services (Wang 2006), a development that could benefit poor and elderly people who may need more information, help and patience than other customers. Post offices have also responded by increasing efforts to promote other goods, including offering savings accounts and selling insurance (interview, Wanzai post office manager, 5 July 2005). Although the banks and credit cooperatives will undoubtedly be able to increase their share of the remittance services market through their new card, a competitive advantage that China Post may nevertheless retain is the relative safety of its service; this is because in China all loss arising from the theft of a bank card is borne by the customer. With the recent dramatic increase in mobile phones in China's rural areas (Murphy 2006) there may in future also be possibilities for instituting linkages between financial institutions/postal agencies and telecommunications companies which in other developing countries enable migrants to use their mobiles to remit their money rapidly and safely (United States Agency for International Development (USAID)/Department for International Development (DFID) 2005).

How are remittances used?

Without detailed records of household budgets it is difficult to gauge exactly how remittances are allocated, or the role of remittances in freeing up alternative sources of household income for different kinds of usage. Even so, a wealth of data based on interviews with rural people about remittance use, field observation and other records suggest that both in China and in other developing countries only a small portion of money is allocated to productive investment, particularly to agriculture, while the vast majority is used for consumptive investment (de Brauw *et al.* 2003; Huang and Pieke 2003; Huang and Zhan 2005b; Murphy 2002). Here 'productive investment' refers to investment in activities that increase the household's capacity to earn money. 'Consumptive investment' refers to goods and services that more immediately improve the wellbeing of the household members. In using the terms 'productive investment' and 'consumptive investment', I follow Rozelle *et al.* (1999b) to indicate that, unlike some policymakers and migration scholars (e.g. Lipton 1980), I recognise that the money used for consumption contributes positively to livelihoods and is therefore not wasted.

The discussion below reviews scholarly findings on the investment of remittances in the productive areas of agricultural production, land purchase and the establishment of small businesses, and in the consumptive areas of housebuilding, the purchase of consumer goods, and the payment of health and education expenses.

Productive investment

Agriculture

As mentioned, scholarly research suggests that in comparison with other areas of remittance usage, agriculture benefits relatively little from remittances. In particular, when yields are examined, it would appear that remittances offer either no or limited compensation for the labour lost from farming. One study based in poorer regions of Sichuan found that remittances did not counteract the effects of labour lost from agriculture, with migrant households commonly producing lower yields than non-migrant households (Li 2001). A different study based on a sample of 585 rural households in northern and central China found no link between migration and productive investment, where productive investment is defined as 'investments in agricultural or non-agricultural activities that enhance the income-earning potential of the household'. The authors did, however, find evidence of a 27 per cent increase in consumption investments (de Brauw and Rozelle 2003). Another study, based on a 1995 survey of 787 rural households (Taylor *et al.* 2003) found that most remittances used for agricultural purposes were directed towards purchasing additional inputs to substitute for the labour lost through the migration of a household member. On account of this usage of remittances, migrant households were able to achieve the *same yields* as non-migrant households. Yet owing to the fact that migrant households needed to spend more money to produce farm yields equivalent to those of their non-migrant counterparts, their *net incomes* from agriculture were *less* than those of non-migrant households. In a further article based on the same survey, the authors report (1) that maize yields *fell* for each family member that left the farm and (2) that remittances only partially offset this migration effect (Rozelle et al. 1999b).

Even though remittances seldom support agricultural production to the extent that the money offsets the effects of the labour lost from farming, remittances may nevertheless support agriculture in other ways. One way is that remittances may help to alleviate capital constraints and provide security in risky agricultural sectors where credit and insurance markets are not developed (Stark and Lucas 1988). Case study evidence supports this contention. A study in rural Jiangxi found that on account of remittances, the poorest households with migrant members no longer needed to rely on borrowing seed and fertiliser from neighbours before the harvest and making their repayments in grain after the harvest (Murphy 2002). Also, observations from rural Chifang prefecture, Inner Mongolia Province, indicated that remittances provided an insurance buffer and saved many households from destitution during periods of ecological disaster (Huang and Zhan 2005a).

In noting the role of remittances in providing money for financing on-farm activities and insuring against on-farm income shocks, Rozelle *et al.* (1999b) suggest that improved rural credit systems would help to limit 'migration-induced reductions to household labour'. This perspective is valuable for highlighting the need for micro-credit in rural areas. It may be argued, however, that the extension of micro-credit in rural areas is most useful as a pro-poor and pro-development project to complement remittances rather than as a mechanism aimed at reducing migration. Indeed, a decade ago Chinese policy researchers (Research Group of the Yichun Prefecture Agricultural Bank 1996) observed that there was scope for mobilising the immense domestic rural savings accrued through remittance deposits towards an expansion in micro-credit. Harnessing remittance deposits could therefore be a way for policy-makers to further their efforts, intensified since 2001, to expand rural micro-credit (Qu 2005: 25).[2]

Scholars have proposed several reasons to explain why people in rural China invest only a small proportion of their remittances in agriculture or other forms of economic production. One factor already alluded to is that the low rate of return to productive capital in poorer areas discourages investment in such activities (de Brauw *et al.* 2003). Another reason is that insecure land rights discourage investment. Indeed, according to Rozelle *et al.* (1999b) the degree of land adjustments within a village – to take into account changing household demographics – negatively affects migrants' inclination to remit. My fieldwork observations lend some weight to the finding that insecure land tenure makes people unwilling to invest in farming. Several farmers told me that on account of insecure land tenure they would not consider using remittances for fixed infrastructure such as tube-wells. I also met people who were leery of contracting additional land, for instance hill land for specialist orchard cultivation, for fear that once the soil was prepared, user rights could be revoked. Yet precisely because most money is used for consumption purposes, it is doubtful that insecure land tenure would affect a willingness to remit *per se*. A further factor that would have discouraged investment in farming during the 1990s, the time of the above mentioned studies by de Brauw *et al.* and Rozelle *et al.*, is the steep agricultural and land-based taxes. Preliminary anecdotal evidence suggests that the removal of rural taxes and the payment of subsidies to farmers for grain production have in large part addressed this problem. During a field visit to an agricultural county in Jiangxi Province in December 2006, I met several farmers who told me that on account of the tax reforms and subsidies, they were willingly renting the land of migrant neighbours at 100 *yuan* per *mu* (0.1647 acres).

A final reason that studies find rural people invest only a small portion of remittances in production is that small-scale labour intensive family farming does not generate a need for sophisticated machinery (Qu 2005: 10). For machinery investments to be viable, it would be necessary for farmers to consolidate their plots into larger entities. There are though some localities in China where extensive investment of migrant earnings in farm machinery has been reported. In such localities, flat terrain, fertile land and a high proportion of out-migration have prompted township governments to coordinate the provisioning of mechanised ploughing and harvesting services. Migrants are willing to pay because such services enable them to either direct the additional income and grain towards the upkeep of their left-behind relatives, or else to gain extra income for themselves (Murphy 2002). Moreover, some rural households have used remittances as well as other sources of money to invest in modest labour-saving equipment suitable for small-scale family farming; for example, a machine popular during the late 1990s could both pump water and thresh grain at harvest time (*choushuiji*) (Murphy 2002). There is also evidence that returned migrants invest substantially more than migrants and non-migrants in items of agricultural machinery such as threshers, water pumps, ploughing machines, seeders, grain processors, and feed processors.[3] Unlike migrants, the attention of returnees has turned to farming and, unlike non-migrants they have urban savings (Zhao 2001).

Land

Most rural migrants in China have access to the use rights of a plot of land for the purposes of agricultural production. This land provides at least a subsistence safety net from which households can pursue livelihood diversification strategies that often include migration. Owing to the fact that under the present contract system, land rights cannot be bought or sold and land leases are not always completely secure, households in rural China do not generally use remittances to buy land. This differs from the situation in the rural-sending areas of some other developing countries. For instance, in rural Cambodia, following the 1995 land reform movement to redistribute land, remittances have enabled some debt-ridden households to hold on to their land or to buy back lost land (Maltoni 2005). As another example, in parts of rural Mexico, property acquisition by migrants has been found to exacerbate inflationary trends in local land markets, adversely affecting the wellbeing of poorer non-migrants and even forcing some tenant farmers to migrate (Massey *et al.* 1987: 236–41).

While on the whole rural households in China have not used remittances to buy farmland, there are nevertheless some instances of returned migrants using the emergence of an invisible land market in the areas surrounding towns and transport routes to gain collateral for their business activities. A house situated along a transport route in a township or on the outskirts of a county seat or small city can be used as security for obtaining loans from rural credit cooperatives. A prerequisite for applying for a loan is obtaining a ' house property certificate' which is issued after an official from the Township Land Bureau has inspected the property. In practice, the geographical location of the land is incorporated into the value of the building so that the building is assessed at a value much higher than bricks and mortar. Remittances are used to pay for the land administration fees and per *mu* (0.0667 hectares) land-use fees needed to obtain such prime building sites. Remittances therefore provide a way for rural people to obtain a desirable property location and by extension, to gain access to local credit. Hence, even though farmland is not really an outlet for remittances in China, a small number of rural people use their urban savings for strategic land investment (Murphy 2002).

Business creation

According to a 1996 survey by the China Rural Development Research Centre, around one-third of migrants from China's interior provinces had returned home (cited in Murphy 2002: 125). Meanwhile, the age structure of migrant labourers in large cities has remained stable since the early 1990s, indicating that many migrants have returned to sending areas (Huang and Zhan 2005a). While the vast majority of returned migrants in China go back to farming, there are nevertheless a portion of migrants who use their urban savings as well as their urban contacts, skills and information to set up businesses. Even though returned migrants who create businesses may not be large in number (Bai and He 2002; Murphy 2002),[4] the fact that a portion of them are entrepreneurial and have urban connections mean that their impact on the local economy may be greater than their numbers alone suggest (Ma 2001; Murphy 2002).

Instances of significant business creation by returned migrants have been reported in counties in a range of interior provinces including Sichuan, Anhui, Shandong and Jiangxi provinces (Ma 2001; Mobility and Rural Development Research Group 1999; Murphy 2002; Zheng 2000). It is worth noting however that most of the counties and prefectures in which returned migrants create businesses are at least of an average economic standing in their provinces and are relatively close to main roads or railways. It is also worth noting that for the most part, entrepreneurial returnees resettle not in the home village but in nearby townships or county seats (Murphy 2002).

Case studies find that in creating businesses, the vast majority of returnees replicate the economic activities of their former urban employers. Sometimes this is a recipe for success. For instance, some returnees make shoes and handbags in a quality and style suitable to the consumer level of the rural market or else they carry out part-processing for urban firms. Sometimes imitating urban ventures is a recipe for failure. In such instances the returnees invest in undertakings entirely unsuited to the economic and consumer environment of the rural townships and county seats.

Research suggests that compared with men, it is harder for returned women to use their urban savings in entrepreneurial ways (Jacka 2006: 153). One reason is that on account of patrilocal residence patterns, women identify less strongly than men with their hometowns. Relatedly, women have less access to official and commercial contacts needed for obtaining access to resources such as credit, land and permits. A final restriction occurs because in rural areas women tend to be seen as less able than men. Even so, a portion of returnees who set up businesses are female, though their businesses tend to be small-scale and domestically oriented, for example hairdressing or sewing.

While it would be an exaggeration to view returnee business creation as an elixir for rural development, it is nevertheless the case that some returned migrants have made contributions to some townships and county seats by setting up businesses. Moreover, in many of these localities, local officials have played an important role in helping the entrepreneurs by creating a positive business environment and by extending practical assistance. The State Council's 2006 comprehensive policy report on migrant workers endorses this phenomenon, reiterating earlier policy guidance that former migrants be encouraged to play a role in building up industrial zones in the interior.

Consumptive investment

Housebuilding

In common with the families of migrants the world over (Fadayomi *et al.* 1992; Lipton 1980; Mills 1997; Oberai and Singh 1980; Rempell and Lobdell 1978), the lion's share of remittances and urban savings are used for house repairs and house construction (e.g. Hunan Labour Transfer and Population Mobility Research Group 1995). Research in Yudu county in Jiangxi province finds that rural households allocate roughly 60 per cent of their remittances to house construction (reported in *Renmin ribao*, 6 February 2006). Such expenditure enables the family members remaining behind to enjoy the benefits of walls that do not let in the wind and ceilings that do not leak. Such improvements in accommodation are valuable in enhancing family wellbeing.

In rural China preparations to build a house have customarily involved mobilising money to buy materials and mobilising social contacts to help with dredging sand and laying bricks. With migration, however, the members of many households have become short of time and have been unable to participate in reciprocal donations of labour for housebuilding. They have instead started to replace labour donations with cash ones. Alongside the inflow of cash into the rural economy and the boom in house construction, in many rural areas local construction teams have emerged. The members of these construction teams, some of them skilled returnees, offer not only their labour but also the ability to provide new styles and standards of decoration. In demonstrating the family's control over social and economic resources, the front of the house and its decoration increasingly correspond with the face of the family (Wilson 1997).

In a society with virilocal marriage patterns, it is nearly impossible for a man to find a bride if he does not have a respectable house to offer. In predominantly agricultural localities, using urban earnings to build a house is therefore essential for rural people to feel that they are respectable in their communities. The ultimate expression of respectability for a man in rural China is taking a bride and continuing the family line. Building a large house is therefore not simply an ostentatious display on the part of migrant households, it is also a prerequisite for meeting the basic human needs of self-respect and full social integration.

A smaller portion of rural people use their urban savings to buy a house in the county seat. This is a way for them to establish an urban lifestyle closer to home. It is also a way for them to ensure that their children are able to attend better quality primary, middle and high schools (*Renmin ribao*, 6 February 2006).

Consumer goods

In the poorest regions, remittances help to pay for everyday items such as soap, matches, batteries and clothes. In middling and richer regions, remittances have sustained rural consumption of manufactured durables. Demand for goods such as television sets, air conditioning units, washing machines and motorcycles have been particularly strong, and the rural demand for these items increased by 17 per cent in 2003 (Kynge 2004). Partly owing to poorer infrastructure, the penetration of white goods into the countryside is only 25 per cent that of the cities (Qu 2005: 30), suggesting considerable scope for continued growth. Remittances help to stimulate the consumption of such durables by generating higher incomes. Moreover, unlike payment for work in farming, which comes only after the harvest, cash from urban wages is more immediate and regular.

Remittances are commonly used by rural people to buy consumer goods as part of the gifts exchanged at lifecycle celebrations; in particular, marriages. For instance, migration enables many young men to accumulate funds for purchasing the modern consumer goods that contribute to a respectable bride price. And migration enables women to earn money to buy some dowry items. In the case of a new bride, a larger dowry enhances her equality in her interdependence with her husband. But on the downside, by lifting the standards of what counts as 'respectable', remittances may increase the total amount of resources necessary for producing a respectable bride price or dowry. In turn, the pressure to keep up can force some people into debt (Murphy 2002).

Health and education

Expenditure on health and education are common uses of remittances in rural areas. This is because in most of rural China medical care is largely privatised and involves the payment of user fees. With regard to education, even though the government has repeatedly stipulated that all children are to receive nine years compulsory education, many poor rural areas lack the money to pay teachers' wages and to run schools, so must charge fees. In addition, parents need to pay for textbooks, uniforms and stationary. Health and education expenses place a strain on household budgets and cannot be covered with income obtained through agriculture alone. This is clear if we consider that in 2003 a rural household's annual per capita income from agriculture was just under 300 *yuan* while its annual per capita expenditure on health and education was 1200 *yuan* (Huang and Zhan 2005b).

In cases of poor households, the need to earn money to buy medicine and to pay for siblings' or children's education may be a primary motive for migration (*Renmin ribao*, 6 February 2006). Indeed, in his survey of 62 rural migrant women, one-fifth of them explicitly told the scholar Zhan Shaohua that they remitted money to pay for their younger siblings' education (Huang and Zhan 2005b).

A survey conducted across 12 townships in rural Shanxi suggests that children may benefit from migration and remittances. Children from migrant households were found to obtain higher test scores than their classmates from non-migrant households (Chen *et al.* 2007). One reason may be that remittances increase the wealth of the household – the general education literature associates higher household income with greater educational attainment (Brooks-Gunn *et al.* 1997; Hannum and Park 2003; Hanson *et al.* 1997). Another reason may be that parental migration may indicate parental commitment to a better future for their children. Further, children's awareness of their parents' sacrifice may impel them to study harder to fulfil parental expectations (Chen *et al.* 2007).

Chinese scholars have expressed the hope that with the eleventh five-year plan's commitment to remove rural taxes, raise rural income and improve rural public goods and services, that rural health and education will become more affordable. This in turn, they suggest, will free rural people to use their remittances for purposes other than meeting basic health and education costs (Huang and Zhan 2005b).

Given that health and education costs and rural tax burdens are heaviest in the poorest agricultural regions where local governments lack an industrial or commercial tax base for funding public goods, if the new policies were fully implemented, then the poverty alleviation impacts of remittances would be most enhanced in these regions.

Who gets the money?

In the view of Chinese planners, the household arranges migration and manages the investment of remittances to the benefit of its members. This assumption can be seen in central government guidelines for poverty alleviation which urge each rural household to have at least one member working in a local off-farm undertaking, a specialised agricultural sideline or as a migrant in a developed region. It can be seen in local government slogans such as 'the migration of one person frees the entire household from poverty'. It is also evident in corresponding prescriptive media reports about rural households which use their remittances to buy fertiliser, set up businesses, invest in the senior high school education of their children and generally progress.

The view that it is the household or the family which harnesses and allocates the benefits of migration is wholly consistent with the pivotal role allocated to the household for managing other aspects of social and economic activity in China, for instance the distribution of welfare entitlements through the household registration system and the management of farmland through the household contract responsibility farming system. Such assumptions about household economics also prevail in international policy and academic circles. For instance, these assumptions underpin the household livelihood diversification and family strategies approaches to migration which are common in Western academic studies and which have in turn been informed by earlier neoclassical economic models in which altruistic household heads and household members cooperate to ensure that everyone benefits from shared resources (Chant and Radcliffe 1992; Kabeer 1994).

Yet while rural family members undoubtedly benefit from the deployment of remittances in production and consumption, this section balances the picture by showing that remittances are not cost free and that the distribution of the human costs and economic benefits of migration is unequal.

Costs to the migrant

As mentioned earlier, migrants incur tremendous human costs in order to remit. Many migrants send home money as soon as they receive their wages so that they will not be tempted to spend more than a stringent allowance on themselves (Yang 2005). The imperative to remit is such that they send home a large portion of their income: Li Qiang's survey of 493 migrants in Beijing found that half remitted at least 40 per cent of their wages and 29 per cent remitted at least 60 per cent (Li 2001).

In remitting such a large portion of their earnings, migrants commonly deny themselves money for warm bedding, clothing and decent food. Migrants also often deny themselves accommodation that is inhabitable (Wang and Wang, this volume). This unwillingness of migrants to invest in their own wellbeing combines with gruelling working hours and poor conditions to inflict a toll on their health. To make matters worse, in the event of sickness the migrants generally feel unable to spend money on prohibitive doctors' fees and medicine. Indeed, the need to remit money is one reason that most migrant workers are reluctant to contribute to enterprise-based social or health insurance schemes in the few places where these are available. Some commentaries in the urban media actually go so far as to criticise rural migrants for placing all their attention on remitting their money instead of helping to stimulate the urban market economy through city-based expenditures (Guang 2003: 629).

Intra-household and intra-family inequalities

Viewing remittances as a tool that promotes collective welfare overlooks intra-household and intra-family inequalities. With regard to the distribution of remittances, there are at least three overlapping dimensions of inequality: the inequality between the migrant and the 'left behinds'; intergenerational inequalities, particularly discrimination against older people; and inequalities along gender lines. Each is considered in turn.

The migrant and the left-behinds

Inequality between the migrant and those left behind has been usefully examined by the geographer Jørgen Carling through the lens of 'information asymmetry' (Carling 2005). 'Information asymmetry' refers to a situation in which one party to a transaction has more or better information than the other party. The metaphor of a transaction is pertinent to rural–urban labour migration. This is because there is an exchange between, on the one hand, the rural family's contribution to the initial costs of migration and provision of livelihood security, and on the other hand, the migrant's remittance of cash income. According to some accounts, in China in 2003, the initial costs to some destination areas could total close to 1,000 *yuan*: this sum included fees for permits to leave the village and verify marital status, fees for permits to reside and to work in the city temporarily, the payment of a deposit to the employer, travel costs and subsistence costs to cover the initial job search period (Thorborg 2006). One survey of remittance behaviour and family ties in China found that the migrants who remitted the most reliably were those who received money from their families to contribute to the initial costs of migration and whose footing in the city was the least secure (Cai 2003).

Information asymmetry plays a role in the security/support remittances trans-action between the migrant and the left-behinds because the migrant has more information than the left-behinds about the amount of money that has been earned. Migrants may wish to use this information asymmetry to keep a larger share of the remittance money for their own use or for use by themselves and a spouse rather than giving a larger share of the money to elderly parents or to other family members. Migrants may therefore arrange for money to be sent back to a separate bank account for their own future use, often for building or renovating a house for the new conjugal unit, rather than remitting all the money into the care of rural family members. In rural Jiangxi I encountered several instances in which married male migrants had remitted money to their parents-in-law for safekeeping, because entrusting the money to their own parents could have potentially caused awkward misunderstandings about who was entitled to borrow or use the money. Their concerns were not entirely unjustified: I also encountered instances in which parents who had been entrusted with buying building materials or coordinating the hiring of builders on behalf of their absent adult children had borrowed some of the money for their own purposes (Murphy 2002).

The point here is that remitting money is not solely an act to promote collective family welfare; it is often a means by which migrants save money for their own purposes, and sometimes migrants maintain the information asymmetry when they feel that other family members might wish to make unwelcome claims on these savings. This is not to deny that most migrants want to help their families or at least to be seen as people of conscience who fulfil their obligations. The information asymmetry is used by some migrants to help them to be seen as a worthy spouse, child or sibling, while also enabling them to keep a larger share of their earnings free from wider familial claims.

Generational inequalities

Many of the tensions over the allocation of remittances and other resources occur between generations, and there is an emerging body of literature which suggests that elderly people, particularly those in poorer households and communities, may be the net losers from migration (Ding 2004; Du and Tu 2000; Hong 2004; Murphy 2002; Murphy 2004; Pang *et al.* 2004). In a growing number of situations it is the labour of elderly people in looking after the migrant's land and caring for grandchildren that makes the migration of adult children socially and economically possible. The extent to which remittances compensate for the increased work burden of elderly people varies across communities, households and individuals. At the community level, there are obviously less options in poorer areas for people to use remittances to hire labour because much of the money goes towards meeting basic needs. Another community level factor pertains to the extent of out-migration: the extent of out-migration affects the number of relatives and neighbours remaining in the village who are available to provide care and company, human needs that remittances cannot buy.

At the household level, on the negative side, some studies find that migration is associated with the earlier division of sons from their parents to form their own nuclear households. Once households have divided they keep separate economic accounts. The earlier division combines with migration to mean that ever-increasing numbers of elderly parents live alone. The consequences of living alone vary depending on the wealth and health of the elderly couple. Some couples have their own savings and so they can continue to maintain a decent standard of living. Some couples are poor. Oftentimes, the only support they receive is an entitlement to retain the grain produced on the adult migrant's land. A study of 151 elderly people with migrant children in rural Zhejiang found that 49 (32.45 per cent) of them rated a lack of money as their second greatest difficulty in life, the first being a lack of care and help. According to the study, this percentage of elderly people facing economic hardship was 26 per cent higher than for those elderly people with an adult child by their side (Ding 2004). As Pang *et al.* (2004) show, even when cash from children is available, many elderly parents are reluctant to make a claim on these earnings: unless they are totally incapacitated, they elect to ' work till they drop'. They use their labour to ensure that their sons can build a nice house, marry and establish themselves economically, thereby being in a better position to provide support when absolute need hits. Indeed many elderly people with whom I spoke said that migration was good because it relieved them of the burden of accumulating the funds for a son's new house and the bride price – a traditional duty of parents being to ensure the continuation of the family line.

At the individual level, many elderly people have lower levels of education than the general rural population and their labour in domestic work and farming tends not to generate immediate cash income. They tend to overlook the value of their own labour contributions to their families, and to see themselves as burdens and of little value. They therefore feel reluctant to demand a greater allocation of resources in compensation for their increased workloads. Elderly people who have lost a spouse, especially elderly women without sons, are the most likely to face destitution on account of migration. I encountered several such widowed women who felt desperate at being forced to go to the hills to collect fern for fuel and to plough the land, especially in localities where the latter is customarily a male task (Murphy 2004).

There are at least two explanations why the hardship that elderly people experience on account of migration is overlooked. One is that, in the case of China, people have a deep impression of their own commitment to the virtues of family values and filial piety, and so there is the widespread assumption that all elderly people receive money and care from their children. Another reason is that, as is the situation the world over, elderly people tend to be 'invisible' or else visible only as 'burdens' (Gorman 1995).

In recent years in China there have been some television commercials that urge adult children to remember their parents. However, policymakers and NGOs need also to look to forms of welfare support which go beyond invoking a family responsibility welfare paradigm (Cook 2002). The suggestion by Pang *et al.* (2004) that China's policy-makers need to recognise the heavy work loads and economic self-reliance of the rural elderly by providing practical support for them in the form of credit, education and information are both timely and urgent. Observations from some other countries such as those in South Asia are also instructive. Studies have found that when income or economic resources accrue directly to elderly people, that their status within their families increases (Gorman 1995; Martin 1990, inter-generational solidarities and transfers strengthen (Barrientos 2003) and the 'burden' stigma is lessened. Indeed Sen (1991) showed long ago that visible income contributions and leverage within the household go hand in hand, an insight which could inform interventions to provide support for China's elderly. At the same time, however, the reality of the declining physical health of some elderly people means that a 'productivity-enhancing' approach to welfare will not be appropriate for all.

Gender

Gender is a final dimension that contributes to intra-familial inequality in the distribution of remittances. Gender inequalities are complex to delineate because gender interacts with a range of other factors such as class, age, marital status, reproductive status, working status, personal attributes and community factors to determine who wins and who loses in the distribution of resources. For purposes of simplicity, here I consider access to the resources generated by migration in relation to the migrants, and then in relation to left-behind children and spouses.

Unlike in other countries where women remit more (Chant and Radcliffe 1992: 17), in China it seems that men probably do (Murphy 2002: 107). Migrant men and women have different opportunities for earning money, with women commonly earning less than men (Fan 2004). Men and women also face different pressures that shape their remitting behaviour. Unmarried men in rural China experience immense pressure to remit as a way of saving for repairing or building a house and accumulating a bride price – only in this way can they prove themselves as eligible candidates in the marriage market. With distortions in sex ratios and the increasing out-migration of young women, the amount of money that men in poor regions need to accumulate for a bride price has been increasing. Married men are also compelled to remit because in the patrilocal and patrilineal family system of rural China the duty of providing care for elderly parents falls mainly to the sons.

That said, Chinese women also remit and want to contribute to the care of their natal families. Many young women contribute their earnings to the building of a new family house, yet they do not receive a share of the family property when it is divided because in rural China only the male siblings receive an allocation – so a daughter's prospects depend on marrying into a good family. Young women also contribute to their parents' daily maintenance. I encountered many parents who would say: 'My daughter is very understanding and knows the hardship of home and always sends us money.' Indeed it has become increasingly common for young women to continue to contribute remittances to their parents' livelihoods even after they have married into their husband's family. These contributions by daughters have in some parts of China led to a new appreciation of their value by parents and have raised the status of daughters more generally (Yan 2003; Gates 1996). Several parents I met expressed the view that while sons can be unreliable, daughters are close to their parents' hearts.

With regard to the distribution of remittances to the left behinds, surveys show that migrants' whose wives remain in the countryside remit a higher portion of their wages and on a more regular basis (Li 2001). In some instances, remittances increase the resources and autonomy of women, and also their responsibilities. This is because women remaining in the countryside may be allocated the money to spend as they see fit. In other instances, women experience little increase in financial autonomy, though also little increase in responsibility. In such cases, they may be sent the money only at specified periods for specified purposes with clear instructions, for instance, for buying fertiliser or bricks or paying school fees. As yet, little research has been conducted into the household allocation of resources and power in circumstances where the wife remits and the husband is 'left behind'. Evidence from the Philippines suggests that gendered divisions and valuations of roles and contributions remain in place and remittances become the way that absent wives and mothers perform their feminine caring roles. Meanwhile, some men may turn to drinking and gambling to escape loneliness and to assert their masculinity (Parreñas 2005). This is an important topic that requires future research.

As for the distribution of remittances to children, wider factors pertaining to the allocation of resources to sons and daughters are also relevant. On the whole, daughter discrimination is deeply embedded within rural Chinese society (Croll 2000), which means that, especially in circumstances of poverty, a greater share of scarce resources goes towards the education and health of sons. In very poor households, migration may also result in the withdrawal of girls from school to help with domestic and farm work. More positively, however, in generating more cash income, remittances may mean that both girls and boys benefit from the increased household wealth, even if in general the latter benefits more.

In sum, prevalent concepts such as 'local development', 'household strategies', 'family decision-making', 'household livelihoods', 'family values' and 'filial piety' obscure inequalities along gender and generational lines in the distribution of the human costs and economic benefits of migration.

Instances of non-remittance

While the vast majority of migrants do remit, a small proportion of migrants are either unwilling or unable to remit. In Li Qiang's surveys, mentioned previously, over one-quarter (29.7%) of 451 migrant households in rural Sichuan did not receive any remittances and around one-quarter (24.7%) of 493 migrants in Beijing did not remit. Three reasons that migrants send only few or no remittances include ill fortune in the cities, the costs of adapting to an urban lifestyle and the wealthy background of some migrants.

To begin with ill fortune in the cities, a common problem faced by millions of Chinese migrants is that of unpaid wage arrears. Bosses have habitually withheld wages as a strategy for retaining labourers and earning interest on the money. The situation became so bad during the Spring Festivals of 2002 and 2003 that large numbers of migrants protested. In response, the State Council issued its

No. 2 Document (2002) and No. 1 Document (2003) demanding fair treatment for migrants and the prompt payment of wages (Huang and Zhan 2005a, 2005b). While such government intervention has yielded significant improvements, the problem of withheld and even unpaid wages nevertheless continues to plague the lives of many migrants and their families.

In other circumstances, failure to remit stems from migrants' increased efforts to integrate into urban society and acquire urban citizenship. Becoming urban involves buying clothes and improving one's image to enable the migrant to shed visible evidence of their rural origins. Becoming urban may also involve paying for city night-school classes. A final set of expenses are incurred as the migrant tries to accumulate funds to buy a house in the city, to resettle other family members and to enrol children in a local urban school. Scholars working in different countries commonly find that long-term settlement in the destination area corresponds with a sharp decline in remittances (Ahlburg and Brown 1998). In the case of China, a comparison of the remitting behaviour of migrants with temporary urban residence permits and migrants with permanent urban residence permits similarly suggests that the former remits more frequently and in larger amounts than the latter (Cai 2003). It is likely that as China's urbanisation proceeds and more people take up urban citizenship, remittances will decline.

A final circumstance associated with few or no remittances is when the migrants come from richer families and richer regions that have many local off-farm activities. These migrants tend to remit less money and less frequently because the family at home does not have such strong need (Li 2001). By contrast, migrants remitting to poor families gain a sense of achievement and improved status in their communities because their remittances have such a dramatic and visible impact on the material lives of their family members. Even in instances of non-remittance in poorer localities, however, provided that the remaining labourers are healthy, migration may still help the family's economic circumstances because the number of people eating from the common rice pot is reduced.

Conclusion

Many people in rural China enjoy access to reliable and affordable services for sending and receiving remittances. The money that migrants send to their families is a way for them to maintain strong bonds of care, affection and belonging in an environment where urban employment and agriculture are both precarious. The remitted money is of huge importance to the welfare of the members of the recipient households, and in poor areas these funds can commonly make up as much as half the total income of a household.

In China, as elsewhere in the world, remittances are used mostly for consumptive investment in housebuilding, the purchase of consumer goods and health and education expenditure. Such consumptive investment improves rural people's material living conditions, sense of self-respect and capacity to participate in their communities. The productive usages of remittances are also important. In some circumstances migrant earnings partially offset the effects of the household labour lost from farming and provide an insurance buffer in the risky agricultural sector. And in some localities returnees use their urban savings to create businesses.

Yet the focus of national level policy-makers, local level officials and scholars on the economic and developmental uses of remittances ignores the immense human costs endured by migrants and their families. It is perhaps for this reason that so far little policy debate has contemplated the kinds of institutional measures that would be required to remove the need for families to endure protracted separation in order to sustain a decent livelihood. The focus of policy-makers, officials and scholars on the role of remittances in the livelihoods of rural households also overlooks inequalities within households and families along gender and generational lines. There is, therefore, scope for both government and non-government agencies to pay greater attention to the needs of those people whose work burdens and emotional hardships increase on account of migration, but whose claim on a share of urban earnings is limited, in particular elderly people. Finally, official and scholarly attention to remittances as a tool for lifting households out of poverty overlooks the reality that many rural households lose labour but do not receive remittances. So, more evidence is needed to assess the linkages, if any, between non-remittance and rural poverty.

Notes

1 A consortium of 31 public and private agencies housed within the World Bank.
2 See Qu (2005). China has the highest national savings rate in the world, but its financial system is incapable of channelling these precious resources to domestic private entrepreneurs. At a macro-level, inefficiencies in turning domestic savings into credit have led to China's heavy reliance on FDI to fund industrialisation and urbanisation, even though there are sufficient domestic reserves for this purpose.
3 A six-province survey of 824 households found that returnees invested 100% more than other rural people in agricultural equipment (Zhao 2001).
4 An article in a 2000 issue of *China Daily* cites Chinese social scientists to say that 4% of some 10 million surplus rural labourers have used their earnings to establish businesses in their hometowns (Zheng 2000).

References

Ahlburg, D.A and Brown, R.P.C. (1998) 'Migrants' intentions to return home and capital transfers: A Study of Tongans and Samoans in Australia', *Journal of Development Studies*, 35(December): 125–151.

Bai N. and He, Y. (2002) 'Returning to the countryside versus continuing to work in the cities: A study of rural-urban migrants and their return to the countryside of China', *Shehuixue yanjiu* (Sociological Research), 6(3): 64–78.

Barrientos, A. (2003) 'What is the impact of non-contributory pensions on poverty? Estimates from Brazil and South Africa', working paper 33, Institute for Development Policy and Management, University of Manchester.

Benjamin, D., Brandt, L. and Giles, J. (2005) 'The evolution of income inequality in rural China', *Economic Development and Cultural Change*, 53(4): 769–824.

Brooks-Gunn J., Duncan, G.J. and Maritato, N. (1997) 'Poor families, poor outcomes: the wellbeing of children and youth', in Duncan, G.J. and Brooks-Gunn, J. (eds) *Consequences of Growing Up Poor.* New York: Russell Sage.

Cai F. (2005) 'Invisible hand and invisible feet: internal migration in China', unpublished manuscript.

Cai, Q. (2003) 'Migrant remittances and family ties: A case study in China', *International Journal of Population Geography,* 9(6): 471–483.

Carling, J. (2005) 'The human dimension of transnationalism: asymmetries of solidarity and frustration', paper presented at the *Migration and Development Within and Across Borders*: Concepts, Methods and Policy Considerations in International and Internal Migration, New York, 17–19 November, Social Science Research Council.

Chant, S. and Radcliffe , S. (1992) 'Migration and development: The importance of gender' in Chant, S and Radcliffe, S. (eds) *Gender and Migration in Developing Countries.* London: Belhaven, pp. 1–29.

Chen, Y. (2006) 'Zhonghang jiang xia da lidu tuidong feixianjin zhifu gongju de shiyong he puji' (The central bank will use great efforts to promote the use and spread of instruments for non-cash payments) *Jinrong shibao* (Financial times), 23 January 2006.

Chen, X., Qiuqiong, H., Rozelle, S. and Linxiu, Z. (2007) 'Migration, money and mother: The effect of migration on children's educational performance in rural China', working paper, Stanford University.

Cheng, E. and Zhong, X. (2005) 'Domestic money transfer services for migrant workers in China', Consultative Group to Assist the Poor (CGAP).

Cooke, S. (2002) 'From rice bowl to safety net: insecurity and social protection during China's transition', *Development Policy Review,* 20(5): 615–635.

Croll, E. (2003) *Endangered Daughters.* London: Routledge.

de Brauw, A. and Rozelle, S. (2003) 'Migration and household investment in rural China', Department of Economics working paper, Williams College, USA, 29 December, available at http://www.williams.edu/Economics/wp/debrauwmiginv_jce_initial.pdf

de la Torre, A. (2005) 'China-labor: Urban workers send $30 billion to rural homes', *Global Information Network,* 13 December.

Ding, Z. 2004 'Guanzhu nongcun liushou jiating renkou liudong dui nongcun liushou laoren de yinxiang' (Paying attention to the families left behind, the impact of population mobility on the elderly left behind in the countryside) *Renkou yanjiu* (Population research) 4.

Du, P. and Tu, P. (2000) 'Population ageing and old age security' in Peng, X. and Guo, Z. (eds) *The Changing Population of China.* Oxford: Blackwell, pp. 77–90.

Fadayomi, T.O, Titilola, S.O., Oni, B. and Fapohunda, O.J. (1992) 'Migrations and development policy in Nigeria' in Toure, M and Fadoyomi, T.O. (eds) *Migrations, Development and Urbanisation Policies in Sub-Saharan Africa.* London: Codesria, pp. 51–111.

Fan, C.C (2004) 'Out to the city and back to the village: the experiences and contributions of rural women migrating from Sichuan and Anhui' in Gaetano, A and Jacka, T. (eds) *On the Move: Women and Rural-to-Urban Migration in Contemporary China.* New York: Columbia University Press, pp. 177–206.

Fu, P. (2005) 'Nongningong jinrong shichang bu rong hushi' (The financial market of migrant workers is difficult to ignore) *Zhongguo jinrong shibao* (China financial times), 29 April.

Gates, H. (1996) *China's Motor: A Thousand Years of Petty Capitalism.* Ithaca: Cornell University Press.

Gorman, M. (1995) 'Older people and development: the last minority', *Development in Practice*, 5(2): 117–127.

Guang, L. (2003) 'Rural taste, urban fashions', *Positions*, 11(3):613–646.

Gu, X. (2006) 'Zhonghang: zhuoyan changyuan li tui yinhangka yingyong fazhan' (Central Bank: taking a long view in strongly promoting the use of bank cards), *Jinrong shibao* (financial times), 7 February 2006.

Hannum, E. and Park, A. (2003) 'Children's educational engagement in rural China', unpublished manuscript.

Hanson, T.L., McLanahan, S. and Thompson, E. (1997) 'Economic resources, parental practices and children's wellbeing', in Duncan, G.J and Brooks-Gunn, J. (eds) *Consequences of Growing Up Poor* New York: Russell Sage, pp. 190–238.

Hare, D. (1999) 'Push versus pull factors in migration outflows and returns: determinants of migration status and spell duration among China's rural population', *Journal of Development Studies* 35 (February): 45–72.

He, L. (1997) 'Lu Gan kaishi kuaihui wangluo, dagongzhe huikuan bu zai shang naojin' (The establishment of an express electronic remittance network between Shanghai and Jiangxi, migrant workers no longer worry about remitting their money) *Xinmin wanbao*, 25 May, p. 22.

Hong, Z. (2004) 'Living alone and the rural elderly: Strategy and agency in post-Mao rural China' in Ikels, C. (ed.) *Filial Piety: Practice and Discourse in Contemporary East Asian Countries*. Stanford: Stanford University Press, pp. 63–87.

Huang, P. and Pieke, F. (2003) 'China migration country study', paper presented at the *Regional Conference on Migration, Development and Pro-Poor Policy Choices in Asia*, Dhaka, Bangladesh, 22–24 June, pp. 36, available at: http://www.livelihoods.org

Huang, P. and Zhan, S. (2005a) 'Internal migration in China: linking it to development' in International Organization for Migration (ed.) *Migration, Development and Poverty Reduction in Asia*. Geneva: IOM Research and Publications Department, pp. 67–84.

Huang, P. and Zhan, S. (2005b) 'Migrant worker remittances and rural development in China', paper presented at Conference on *Migration and Development Within and Across Borders – Concepts, Methods and Policy Considerations in International and Internal Migration*, New York: Social Science Research Council, 17–19 November 2005.

Hunan Research Group on the Transfer of Labour and Population Mobility (1995) 'Hunan sheng laodongli de zhuanhua yu renkou liudong' (The transfer of labour in Hunan and population mobility) *Shehuixue Yanjiu* (Sociological Research), 3: 75–85.

Jacka, T. (2006) *Rural Women in Urban China*. Armonk: ME Sharpe.

Kabeer, N. (1994) *Reversed Realities: Gender Hierarchies in Gender and Development Thought*. London: Verso.

Kynge, J. (2004) 'China's urban workforce fuels rural economy.' *Financial Times*, February 26.

Li, Q. (2001) 'Zhongguo waichu nongmingong ji qi huikuan zhi yanjiu' (Research on China's rural out-migrants and their remittances) *Shehuixue yanjiu* (Sociological Research), 4: 64–76. Available at: http://www.usc.cuhk.edu.hk

Li, S. (1999) 'Effects of labor out-migration on income growth and inequality in rural China', *Development and Society*, 28(1): 93–114.

Lipton, M. (1980) 'Migration from rural areas in poor countries: The impact on rural productivity and income distribution' *World Development*, 8(1): 1–24.

Ma, Z. (2001) 'Urban labour-force experience as a determinant of rural occupational change: Evidence from recent urban–rural migration in China', *Environment and Planning A*, 33: 237–255.

Maltoni, B. (2005) 'Country case study: Cambodia', paper presented at the Regional Conference on Migration and Development in Asia, Lanzhou, China, 14–16 March, 2005.

Martin, L.G. (1990) 'The status of South Asia's growing elderly population', *Journal of Cross-Cultural Gerontology*, 5(2): 93–117.

Massey, D.S, Alarcón, R. Durand, J. and González, H. (1987) *Return to Aztlan: The Social Process of International Migration from Western Mexico*. Berkeley: University of California Press.

Mills, M.B. (1997) 'Contesting the margins of modernity: women, migration and consumption in Thailand' *American Ethnologist*, 24(1): 37–61.

Mobility and Rural Development Research Group (1999) 'Nongmingong huiliu yu xiangcun fazhan: dui Shandongsheng Huantai xian 10 cun 737 min huixiang nongmingong de diaocha' (The return of migrant workers and rural development: a survey of 737 returned rural migrants in Huantai county, Shandong Province), *Zhongguo nongcun jingji* (Chinese Rural Economy), 10: 63–67.

Murphy, R. (2002) *How Migrant Labor is Changing Rural China*. New York: Cambridge University Press.

Murphy, R. (2004) 'The impact of labour migration on the well-being and agency of rural Chinese women: Cultural and economic context and the life-course', in Gaetano, A. and Jacka, T. (eds) *On the Move: Women and Rural–Urban Migration in Contemporary China*. New York: Columbia University Press, pp. 227–262.

Murphy, R. (2006) 'Overcoming the digital divide? ICTs and informationalism in rural China', presented at Rethinking the Rural–Urban Cleavage in Contemporary China, Fairbank Centre, Harvard University, 6–8 October.

Ngai, P. (2005) *Made in China: Women Factory Workers in a Global Workplace*, Durham: Duke University Press.

Oberai, A.S. and Singh, H. (1980) 'Migration, remittances and rural development: findings of a case study in the Indian Punjab' *International Labour Review*, 119(2): 229–241.

Pang, L., de Brauw, A. and Rozelle, S. (2004) 'Working until you drop: The elderly of rural China', *China Journal*, 52(July): 73–96.

Parreñas, R.S. (2005) *Children of Global Migration: Transnational Families and Gendered Woes*. Stanford: Stanford University Press.

Qu, H. (2005) *The Great Migration: How China's 200 Million Surplus Workers Will Change the Economy Forever*. London: HSBC Global Research.

Rempel, H. and Lobdell, R.A. (1978) 'The role of urban–rural remittance in rural development' *Journal of Development Studies*, 14(3): 324–342.

Research Group of the Yichun Prefecture Agricultural Bank Planning Research Group and Zhangshu City Agricultural Bank Planning and Science Federation (1996) 'Laowu shuchu dui nongcun jinrong de yingxiang' (The impact of labour export on rural finance) *Nongcun shehui jingji xuekan* (Journal of rural society and economy), 3: 40–43.

Riskin, C. (2007) 'Has China reached the top of the Kuznet's curve?' in *Paying for Progress in China: Public Finance, Human Welfare and Changing Patterns of Inequality*. London: Routledge, pp. 29–45.

Rozelle, S., Guo, L., Shen, M., Hughart, A. and Giles, J. (1999a) 'Leaving China's farms: Survey results of new paths and remaining hurdles to rural migration' *The China Quarterly*, 158(June): 367–393.

Rozelle, S., Taylor, J.E. and de Brauw, A. (1999b) 'Migration, remittances and agricultural productivity in China', *The American Economic Review*, 8 (2): 287–291.

Sen, A. (1991) 'Gender and cooperative conflicts' in Irene Tinker, I. (ed.) *Persistent Inequalities: Gender and World Development.* Oxford: Oxford University Press, pp. 123–149.

Stark O. and Lucas, R.E.B (1988) 'Migration, remittances and the family', *Economic Development and Cultural Change*, 36(April): 465–482.

State Council (2006) 'The State Council Directives on Matters of Migrant Workers', State Council, Beijing, 28 March.

State Council Office (2002) 'State Council Document No. 2: Improving Management and Services for Rural Labour Migrants, Beijing', summary presented in Huang Ping and Frank Pieke (2003) 'China migration country study', Regional Conference on Migration, Development and Pro-Poor Policy Choices in Asia, 22–24 June, 2003, Dhaka, Bangladesh, p. 36, available at: http://www.livelihoods.org

State Council Office (2003) 'State Council Document No. 1: Improving Management and Services for Rural Labour Migrants, Beijing', summary presented in Huang Ping and Frank Pieke (2003) 'China migration country study', Regional Conference on Migration, Development and Pro-Poor Policy Choices in Asia, 22–24 June, 2003, Dhaka, Bangladesh, p. 36, available at http://www.livelihoods.org

Tao, R. and Xu, Z. (2007) 'Urbanization, rural land system and social security for migrants in China', *Journal of Development Studies*, 43(7): 1301–1320.

Taylor, J.E., Rozelle, S. and de Brauw, A. (2003) 'Migration and incomes in source communities: A new economics of migration perspective from China', *Economic Development and Cultural Change*, 52(1): 75–101.

Thorborg, M. (2006) 'Chinese workers and labor conditions from state industry to globalized factories: How to stop the race to the bottom', *Annual New York Academy of Sciences*, No. 1076: 893–910.

United Nations Development Programme (2005) 'The potential role of remittances in achieving the millennium development goals – an exploration', UNDP Background Note, 10 October.

United States Agency for International Development/Department for International Development (2005) *Migrant Remittances*, 2(1) (April).

Wang, F. (2002) 'Nongmin zengshou nanti shentao – jianlun mingongchao' *Hongguan jingji yanjiu* (Macro-economic research), 3: 20–26.

Wang, G. (2006) 'Cong yinhang kan ka shihang kan lüka yewu fazhan' (Looking at developing the business of the green card in the card market). Available at http://www.chinapostnews.com.cn/854/08540601.htm

Wang, S. (2003) 'Zai nijing zhong boji – feidian yiqing dui youzhengju de yingxiang yiji cuoshi' (Battling in adverse circumstances – the impact of SARS on the Post Office and its Countermeasures), 27 June. Available at: http://www.chinapostnews.com.cn.

Wilson, S. (1997) 'The cash nexus and social networks: mutual aid and gifts in contemporary Shanghai villages', *China Journal*, 37(January): 91–112.

Yan, Y. (2003) *Private Life Under Socialism.* Stanford: Stanford University Press.

Yang, L. (2005) 'Guanzhu chengli de nongmingong: zhengqian bu gan hua, quan wei butie jiali yong' (Attention to the rural migrant workers in the cities: they don't spend their earnings, they send it all to subsidize their families), *Gongan wang* (Public Security Net), 3 August. Available at: http://law.anhuinews.com

Zhan, S. (2004) 'Zuzhi nongmingong de shehui paichi: Zhongguo nongcun fazhan de xin jiaodu' (Preventing the social exclusion of rural migrant workers: A new perspective on the development of the Chinese countryside). Available at: http://social-policy.info/912.htm

Zhao, Y. (2001) 'Causes and consequences of return migration: Recent evidence from China', 30 November. Available at http://ccer.pku.edu.cn/download/475-1.doc

Zheng, Y. (2000) 'Migrants help, so help migrants', *China Daily*, 2 November. Available at: http://chinadaily.com.cn

3   *Hukou* reform and social security for migrant workers in China

Ran Tao

Introduction

China, the fastest growing economy in the world, is currently in the middle of its drive to urbanize. The main objective of urbanization is to encourage a shift in the population from rural to urban areas and from rural-based agricultural labour to mainly urban manufacturing and service sectors. Given China's legacy of a planned economic system that limited the expansion of cities and rural to urban mobility, sustainable urbanization is particularly important in China.

Though economic reforms since the late 1970s have led to the gradual relaxation of restrictions on labour mobility originally intended to prevent urban unemployment, the old institutional arrangement of the Household Registration System (henceforth, *hukou*) has still not been fundamentally reformed.[1] *Hukou* remains an obstacle to permanent rural–urban migration and to movements across administrative regions (Au and Henderson 2002).When farmers move to the city to look for work, they are still mostly treated as second-class citizens, even when they find productive employment. Urban officials place strict limits on permanent migration from rural to urban areas and across regions. Rural migrant workers are denied equal access to social security benefits, housing subsidies and quality education in urban public schools for their children available to those officially classified by the *hukou* system as 'urban residents' (Wang 2004; Tao and Xu 2007).

At present, migrant workers are primarily found in temporary low-income jobs in cities. For most, labour protection is minimal and wage arrears and lack of social insurance coverage is common. The lack of social insurance for migrant workers is not only the result of the relatively low-income and temporary nature of their employment, but also the lack of an institutional framework with appropriate social insurance schemes catering to their needs.

This essay analyses how China's present *hukou* system and social security system have affected the livelihoods of the country's growing migrating labour force. It is argued that with economic transition, an urban *hukou* is now mostly associated with urban-based social assistance, housing security and public school services, while the social insurance coverage is increasingly linked to employment.

Therefore, future policy changes not only warrant breakthroughs in *hukou* reform, but also call for a well-designed institutional social insurance framework appropriate for migrant workers.

The rest of the chapter is organized as follows: the second part analyses China's evolving *hukou* system and its impact on the country's labour mobility patterns. The third part describes social security arrangements for city-based migrants. The fourth part explores the implications of China as a large developing country in economic transition for its *hukou* system reform and the provision of social security for migrant workers. The fifth and final part argues for a coordinated policy framework to address the challenges in China's future *hukou* and social security reforms.

China's evolving *hukou* system and labour mobility

During the central planning period, China adopted a development strategy that prioritized capital-intensive heavy industrial growth. Such a strategy was inefficient because it did not make full use of the country's cheap labour force. To maximize resource mobilization for its capital-intensive industrialization, a system of central planning was established in which a free market for capital, labour and other factors of production was forbidden. Resource mobility across regions and sectors was severely restricted through an integrated set of institutional arrangements (Lin *et al.* 2003). The arrangements specifically targeting labour mobility included a household registration (or *hukou*) system, that required individuals to register with local authorities to gain residence, access to a rationing system that allowed people to buy food, basic necessities and major consumer goods only at the place of their household registration, and inclusion in an urban employment and social security system aimed only at urban workers. An urban *hukou* was therefore associated not only with employment and related welfare provided through work units, but also with food and basic necessities. Such arrangements made it almost impossible for a rural individual without an urban residence permit to move to the cities. For example, without the coupons allocated by the state to urban residents, no food of any kind could be bought at official prices.

As small farmers were effectively excluded from the urban sector during the central planning period, China achieved almost full urban employment. The urban population, less than 20 per cent of the national total, was able to enjoy a wide range of welfare benefits from subsidized food, housing and healthcare, to pensions provided through state work units. In rural areas, the labour force was governed by a system of rural collectives. The rural collectives received production targets from the planning authorities and delivered procurements at low, officially set prices (Fleisher and Yang 2003). They were also responsible for organizing rural labourers for agricultural production and allocating resources among the rural work force. Rural labourers received work points for their work, redeemable partly in grain and partly in cash (Lin 1990). Therefore, both the rural and urban labour forces were heavily controlled during the central planning period. Rural to urban migration, or even migration between cities, was unlikely without approval from the state agency.

Since the late 1970s, China initiated market-oriented reforms that gradually relaxed restrictions on labour mobility across sectors and regions. Significant policy changes first took place in the countryside as rural industries boomed in the 1980s. In 1985, a key policy document issued by the central government permitted farmers to work and establish businesses in nearby towns, provided they could prove they had the financial means and their own food grain. Significant changes in government labour and industrial policies paved the way for the fast growth of rural industries. In the 1980s, an increasing number of farmers began to seek off-farm employment in Township and Village Enterprises (TVEs hereafter) and rural labour markets began to develop rapidly (Putterman 1992; de Brauw *et al.* 2004). By the late 1990s, TVEs already employed 92.7 million rural workers, and 128.6 million by 1995 (National Bureau of Statistics (NBS), *China Labour Statistical Yearbook*, various years).

Rural–urban migration began to accelerate as China's reforms started to affect the urban sector in the mid-1980s. During this period, the creation and development of special economic zones, the expansion of the non-state sector and the easing of the urban employment policies created a demand for migrants (Meng and Zhang 2001; Cai 2000). Between 1984 and 1988, the Chinese government started to allow farmers to enter cities on condition that they provided their own food (Huang and Pieke 2003). The food rationing/coupon system was gradually dismantled from the mid-1980s onwards, and individuals could buy food at market prices. This meant that rural migrants could now survive in urban areas if they could find employment. As a result, the flow of rural migrants to cities began to increase (Meng and Zhang 2001). By 1988, there were already about 25 million migrants working in urban areas.

However, large-scale rural–urban migration did not occur until the 1990s, when economic growth in the cities took off and further policy changes in urban labour markets were implemented. It was only in this period that the central and local governments began to explicitly encourage temporary migration so that cities could benefit from cheap workers from rural areas. Since 1992, the state has taken various steps to encourage labour mobility between rural and urban areas. The most significant change was the introduction of two special types of residential registration, namely the so-called temporary residential permit and the blue-stamp *hukou* or blue card. Unlike the regular *hukou*, these are not administered by the central government; instead, the design and implementation are up to local governments. While the temporary resident permit can be issued to anyone who has a legitimate job or business in the city, the blue-stamp *hukou* is issued to investors, buyers of property, and professionals.[2] In 1998, the Ministry of Public Security issued new regulations that relaxed controls under the household registration system, allowing those who joined their parents, spouses and children in cities to obtain an urban *hukou*. After years of local experimentation, the Ministry of Public Security started *hukou* reform in small towns and cities in 2001. In most small towns, the minimum requirement for obtaining the local *hukou* was a permanent source of income and legal housing in the locality. Some medium-sized cities also began to relax permanent migration restrictions.

With the introduction of new policies, large numbers of rural migrants started to work in temporary, low-end urban jobs (Chan and Zhang 1999).[3] Rural–urban migration replaced other off-farm activities as the fastest growing segment in the off-farm labour market in the 1990s (de Brauw *et al*. 2002). By the mid-1990s, rural migrants already accounted for about 18 per cent of the total rural, and 34 per cent of the total urban labour force. The Fifth National Population Census indicated that by 2000 there were already 121 million migrants (defined as individuals who had migrated for at least six months in the past year) in China, of which 90 million were found in urban areas, and 88.4 million originated from rural areas (National Bureau of Statistics 2002).

As economic reforms progressed during the 1980s and 1990s, some of the urban *hukou*-related privileges, such as legal residence and commodity rationing, gradually eroded as markets for urban houses and necessities developed. The cities also witnessed the fast expansion of the urban private sector. All these developments enabled rural migrants to make a living in the cities. From the mid-1990s, China witnessed the significant restructuring of state-owned sectors and the gradual development of an urban social insurance system. The traditional work-unit system was gradually eroded and most urban workers with permanent urban residence permits began to obtain their medical insurance, unemployment insurance and pension through their employers. Because these social insurance schemes became increasingly associated with market-based employment rather than the urban *hukou*, a small number of migrant workers who were able to enter urban formal sectors were also covered.

Nevertheless, urban permanent residents today still receive some municipal welfare privileges unavailable to rural migrant workers. At present, the privileges mainly include social assistance (in China's case, the so-called 'Minimum Livelihood Guarantee Scheme' (MLGS) or *dibao*) and certain forms of housing subsidy through locally funded public housing schemes. Local residents also have access to some, albeit declining, employment opportunities from local public sectors such as civil services. Moreover, the children of migrants usually have no, or only very costly access to urban public schools, while the children of permanent urban residents benefit from heavily subsidized schooling (Kwong 2004). In sum, though the urban *hukou*-linked benefits have been declining as China's economic reforms have progressed, the *hukou* system still limits some public services to urban residents with *hukou*.

With the *hukou* system still functioning, China's rural–urban labour mobility pattern is rather unique, with a massive floating population primarily engaged in temporary urban jobs, while in most other developing countries permanent and family migration of rural residents to urban areas has played a central role in the process of urbanization (Yang and Zhou 1999) According to the National Bureau of Statistics (2002), less than 10 per cent of rural–urban migrants in China move with their families. Although some richer migrants have purchased an urban *hukou* and settled permanently in cities, the overwhelming majority still work in cities without permanent urban residence permits.

For most rural migrants the lack of an urban *hukou* has created significant economic and social problems in cities as well as in rural areas. Migrant workers are heavily discriminated against in cities. Lack of social security, lack of regular and 'legal' housing, and unequal access to public schooling all pose serious challenges for rural migrants in their daily lives. Most migrants in cities suffer from the separation from their family, while their parents and children are left behind in rural areas. As more working-age people migrate to the cities, the role of the traditional rural family in providing care for the elderly is eroding and many old people lack care (Murphy 2004). Over 22.9 million children have to attend local rural schools without their studies being monitored by parents who are working in cities, negatively affecting their academic performance (Sui 2005; Ministry of Education 2006).

The *hukou* system has also constrained the progress of China's urbanization (Au and Henderson 2002). Though the official urbanization rate already reached 40.5 per cent by 2003, China's 'urban' population in the official statistics includes not only the urban residents with an urban *hukou*, but also around 90–95 million migrants (and a further 20–25 million dispossessed farmers whose land has been requisitioned in urban expansion). These groups have no permanent urban residence permits and are thus denied access to the public services associated with an urban *hukou* (National Bureau of Statistics 2004; Han 2005).

Social security for migrant workers

When discussing social security for migrant workers, a key distinction must be made between social insurance and social assistance. In social insurance programmes such as unemployment, work-related injury, medical insurance and pensions, beneficiaries (and/or their employers) contribute a designated amount to be eligible to receive corresponding programme benefits, while social assistance programmes are often funded by governments at different levels and programme benefits are needs-based and delivered on grounds of eligibility. Therefore, the roles of government in social insurance and social assistance programmes are quite different. As argued earlier, an urban *hukou* today mainly entails social assistance, housing security programmes and public school services provided through city governments. For migrant workers, the provision of social insurance is more related to the nature of their employment

Temporary jobs and social insurance programmes for migrant workers

Temporary low-income employment and social insurance

During the late 1990s, significant occupational stratification became institutionalized. Many municipal governments implemented regulations to protect urban workers by reserving specific job categories for them and by explicitly advocating that urban residents not be underpaid compared to outsiders.[4] Studies of urban labour markets in the 1990s and early 2000s usually found significant segregation between rural migrants and urban residents in China's urban sector (Zhao 2002).[5] It was common for the floating population to perform jobs that urban workers refused to do (Yang and Guo 1996; Feng *et al.* 2002).

With further liberalization of urban labour markets in recent years, many of the discriminatory employment policies against migrant workers have been gradually removed. In cases where such policies still exist, they are not followed very closely because they would only undermine the economic interests of urban employers. Nevertheless, owing to their temporary and low-income employment, most rural migrants in China are still not covered by any social insurance scheme. Migrants often take jobs that are dirty, dangerous and demeaning (the three 'Ds'), common in industries such as construction and mining for men and sanitation and textiles for women. According to the Ministry of Labour and Social Security (MOLSS) (2006), in less-developed regions such as Anhui and Henan provinces, the average wage of migrant workers was around half that of their urban peers, while in more developed regions such as Guangdong and Zhejiang, the differences were even larger. Given their low wages, many migrant farmers have no incentive to pay for social insurance.

Lack of social insurance for migrant workers is related to their lack of formal labour contracts. A survey carried out in 2004 by the MOLSS on migrant workers in 40 cities across China found that only 12.5 per cent of all migrant workers surveyed had signed labour contracts with their employers. In the construction sector, where rural workers are recruited by private contractors, those recruited are often not even registered for urban temporary residence (State Council Research Department 2006). Lack of labour contracts makes basic labour protection and timely payment of wages difficult, not to mention social insurance coverage. Even in the state-owned enterprises or urban public sectors, where labour protection and benefits are presumably better, migrant workers are usually afforded little protection or benefits. Protection is even lower in non-public sectors, though legally all enterprises should pay into such social insurance schemes for all their employees (Solinger 1999).[6] A survey of large manufacturing enterprises in Nanjing municipality, the capital of Jiangsu province on the country's east coast, found that welfare benefits for workers, above and beyond earnings, for the years 1994–2001 averaged 36 per cent of the earnings in urban state-owned manufacturing enterprises, but only 16 per cent of the earnings in large manufacturing TVEs in counties under Nanjing's administration (Banister 2005).

In many instances, the temporary migrant workers without urban residence permits are left off the books entirely, at least in relation to what is reported to the authorities, and their employment is kept informal. Neither the workers nor their earnings, which are paid in cash, are reported. This means that the employee can avoid paying income tax and any required social insurance deductions, while the employer can avoid paying the required social insurance for the employee. Even when employment is reported to authorities, both employers and employees tend to collude to minimize reported earnings (Banister 2005).

Pensions

The Chinese government has made considerable efforts to establish a unified pension system across the nation and to extend coverage to the private sector and migrant workers. However, such efforts have had very limited success so far (MOLSS 2006).

There are many difficulties in expanding the pension system to include migrant workers who are mostly in informal sectors (Zhao and Xu 2002). According to the state regulations promulgated in 1997, both employers and employees should pay into a mandatory savings account (the 'individual account'), while employers also contribute to a social pooling account (the 'basic account'). Rural workers are no exception. Nevertheless, the level of contributions, together with the prevailing labour market conditions, made it almost impossible for rural workers to be adequately covered. According to state regulations on individual accounts, a worker should pay eight per cent of his or her salary into the individual account and, until very recently, the employer had to pay three per cent of the salary to the individual account once the worker had paid his or her contribution. Moreover, the employer has to contribute another 20 per cent of the total payroll to the basic account, the objective being to pay older employees' pensions out of the basic accounts, saving individual account funds for younger employees' pensions when they retire. This system places a heavy financial burden on employers and many employers have reduced their contributions or stopped them entirely. According to the Ministry of Labour and Social Security (MOLSS 2006), at present only 15 per cent of migrant workers participate in pension schemes.

The main challenge to pension schemes for migrant workers stems from the temporary nature of their employment and the technical difficulties this creates to coordinate pension schemes across localities. The current regulation requires that pension contributions are paid for 15 years before receiving benefits. However, migrant workers seldom stay in one job for a very long time. If a migrant worker participates in a pension programme in one place for several years paying monthly premiums, but then moves to another place, the retiree will not be able to receive the full monthly pension after reaching the retirement age. To address this issue, the present pension programme is to enable the migrant worker to cash in the social insurance contributions from his or her individual account once that worker decides to move on. It is very common for migrant workers who initially participate in a pension scheme in a locality to exit the programme when they move to jobs elsewhere. In Guangdong province, the exit rate for migrant workers is at present as high as 95 per cent. However, cashing-in the individual account implies that the migrant worker will lose the potential benefits from the basic account after retirement age.[7] Given that employers pay the social pooling part, high exit rates also dampen their incentive to contribute in the first place (MOLSS 2006).[8]

High exit rates of migrant workers also reveal another weakness in China's present pension system, which is pooled at the city or county level. Coordinating schemes between cities and counties is technically very difficult since it involves highly complex and easily controversial transfers of pension funds (especially for the basic account) across localities. This difficulty becomes obvious when considering the huge heterogeneity in pension fund endowments and demographic structures across localities. As a result, the current regime tends to result in a situation in which local governments are willing to collect money from migrant workers, but are not eager to use the collected contributions to spend on those who are no longer within their administrative jurisdiction.

Unemployment insurance

In 1999, the Chinese government issued 'Regulations on Unemployment Insurance'. In principle, all staff and workers of urban enterprises (i.e. state-owned enterprises or SOEs), urban collective enterprises, foreign-funded enterprises, urban private enterprises and other urban enterprises) should be covered by unemployment insurance. Given that unemployment insurance has been pooled at city or county level from the start, the centre leaves the actual policy design to provincial governments. In most provinces, separate regulations on unemployment insurance are set for urban permanent workers and migrant workers. For the former, employers usually need to pay two per cent of their total payroll and the individual contribution rate is one per cent of the wage. The programme provides subsidies for up to two years depending on how long the worker and/or the work unit have participated in the unemployment insurance programme.[9]

Unemployment insurance premiums for migrant workers in enterprises in urban areas are paid by the employers. Workers who have worked continuously for one year, or who do not renew their contracts upon expiration or who terminate their contracts before they expire, can apply for unemployment insurance. The benefit is a living allowance in the form of a lump sum depending on the length of time worked. It is up to the province to decide the terms of the contributions and benefits. For example, in Guangdong province, the subsidy is set at 12 per cent of the average monthly salary in the past year for migrants who have participated in the unemployment scheme for one year, with the rate increasing by one percentage point for every additional month in the scheme. In Zhejiang province, migrant workers obtain a lump sum of at least 40 per cent of the total unemployment insurance benefits paid to urban workers (Liu 2006).

However, in practice many rural workers working in cities and township and village enterprises (TVEs) are excluded from unemployment insurance. According to MOLSS (2006), by 2004 the share of migrant workers covered by local unemployment insurance was below 20 per cent.

Work-related injury insurance

The issue of work-related injury has become extremely important in China, as accidents at work take the lives of more than 100,000 people every year and injure another 700,000. Migrant workers account for the majority of those injured or killed in work-related accidents as they often have to take the more dangerous jobs. In the coal mining industry alone over 6,000 deaths occur annually, and this trend has worsened in recent years as the number of China's occupational accidents has soared.[10]

In the past, migrant workers who suffered occupational injuries often found themselves in a hopeless situation as their employers refused to pay compensation for severe work injuries, occupational diseases or death. In the late 1990s, China launched its pilot occupational injury insurance schemes. After a number of trials in several cities, the State Council promulgated the 'Regulations on Workplace Injury Insurance' in 2003, which came into effect in January 2004. By the end of June 2004 as many as 49.96 million employees had underwritten this insurance scheme (MOLSS 2006).

Under the current scheme, all employers are required to take out industrial injury insurance for their employees. The insurance covers various types of injuries, including casualties suffered during business trips, vehicle accident injuries occurring on their way to or from work, as well as injuries during emergency rescue efforts undertaken to protect state or public interests. The provincial government determines the various premium rates according to the risk of work-related injuries involved in different sectors, and sets several rates within each sector according to the insurance payments and occurrence of such injuries. The benefits mainly include medical costs for work-related injuries; injury and disability benefits, allowance and nursing fees according to the degree of inability to work; funeral payments, pensions for family members and a lump-sum death subsidy, all of which go to the immediate family members of the deceased worker in the case of death resulting from a work-related accident.

At present, the work-related injury insurance scheme is a social pooling programme at the municipal or prefecture level. For example, in Beijing, interim measures stipulate that all migrant workers with formal labour contracts must participate in the programmes. This regulation requires employers to pay an average of 2.6 per cent of total wages for migrant workers, though the rate differs by sector according to the particular risk of workplace injury (Beijing Government Research Group 2006).

However, many migrant farmers are not covered. This is especially true in many urban private and rural township enterprises that do not sign formal labour contracts with their employees. In some extreme circumstances, the employer and employees sign private contracts by which very low compensation is offered in case of workplace injury or death. According to a survey carried out by the Ministry of Agriculture in 2005, only 12.9 per cent of migrant farmers are at present covered by workplace injury insurance (State Council Research Department 2006). Even in Guangdong province with a relatively high coverage of this insurance, only 20 per cent of migrant workers are covered (Liu 2006).

Medical insurance

As in the case of workplace injury insurance schemes, the introduction of medical insurance for migrant workers is very recent and the progress is gradual. In 1998, on the basis of previous trials, the Chinese government promulgated the 'Decision on Establishing a Basic Medical Insurance System for Urban Employees'. Since then, medical insurance has also been gradually introduced in different cities to cover medical costs for seriously ill workers. Again, the insurance policies vary between urban employees with permanent residence permits and migrant workers, and between different localities.

The basic medical insurance funds are pooled at prefecture or municipal level. For urban employees with an urban *hukou*, employers pay around six per cent of the total payroll, and employees two per cent of their wage. Part of the contribution paid by the employer is pooled in a special fund and the rest is paid into the individual account. All of the individual contribution goes into the individual account. The pooling fund is mainly used to pay for hospitalization and outpatient services in case of some chronic diseases, with a standard starting and ceiling amount. The individual account fund is used for general outpatient services.

Different localities set their own policies for migrant workers. For example, in northern China's Jilin province, it is employers who pay the entire contributions and individual migrants do not need to pay. The contributions go to a social pooling account and can only be used for fees incurred for hospitalization. A more recent pilot medical insurance project was introduced in Shenzhen in Guangdong province in March 2005. Under this scheme, migrant workers can apply for reimbursement of up to CNY 60,000 (US$ 7,500) of medical expenses under the system. To be insured, a worker is required to pay CNY 4 (50 US cents) with the employer paying another CNY 8 (US$ 1) every month. This programme already covers 1.24 million migrant workers after one year of operation, and is to be extended to cover 3 million by the end of 2006 (Xinhua News Agency, 26 February 2006).

In general, though, progress in medical insurance coverage for migrant workers is still limited. According to a recent report by MOLSS (2006), only 10 per cent of migrant workers in China are covered by a medical insurance programme.

Hukou-related social assistance and public services

In China, the main urban social assistance programme is the Minimum Livelihood Guarantee Scheme (*dibao*). This programme targets households with a per capita income below the designated urban poverty lines. By the end of 1998, the programme covered many cities across China. It is mostly financed through local budgets, and a central subsidy to poor regions was introduced in 1999 (Zhang 2003). By the end of 2003, 22.47 million urban low-income residents received assistance under this programme amounting to some CNY 58 (US$ 7.3) per capita per month (Ministry of Civil Affairs 2004).[11]

State-sponsored housing security programmes in China take two main forms: one is the so-called Economical Housing Programme in which low-income urban households purchase housing at subsidized prices and are exempt from land-use fees. The other is the Public Housing Programme in which beneficiaries either receive housing at a subsidized rent or a cash subsidy to rent accommodation on the housing market (Pin and Chen 2004).

Though China has made significant efforts in recent years to provide social assistance and housing security for urban residents on a large scale, most rural migrants remain excluded. In addition, it is still difficult for migrant children to gain access to urban public schools. According to a recent survey carried out by the Ministry of Education (2006), in most cities migrant workers have to pay an extra annual fee of CNY 1,200–1,600 (US$ 150–200) for their children to attend urban primary schools, and as much as CNY 2000 (US$ 250) per year for junior middle school. This has contributed to a nine per cent drop-out rate among children who migrate with their parents to cities (State Council Research Department 2006).

Dismantling *hukou* and providing social security for migrants in a large developing country in transition

As a large developing country in economic transition, China's urbanization and economic development are occurring within the context of globalization. China's size, development stage and economic transition and the international environment have important implications for future policy changes in *hukou* and social security for migrant workers.

As a consequence of the general economic globalization process, China's migration is closely associated with the country's redistribution of non-agricultural sectors, with coastal regions experiencing faster growth of foreign investment, employment and exports than interior areas. As industries become more concentrated in coastal regions and cities, migrants also tend to move there. In 1982, coastal regions only hosted some 38.4 per cent of all migrants in China; by 2000 this number had jumped to 64.5 per cent, with a strong regional concentration in the Yangtze River Delta, the Pearl River Delta and northern China's coastal cities. If in 1982 less than five per cent of migrant workers were in the coastal province of Guangdong, their share rose to 15 per cent by 1990, and 27 per cent by 2000 (Li 2003). Such regional migration patterns indicate that the coastal provinces and cities are facing major challenges in *hukou* reform and the extension of social security to migrant workers.

The globalization of the Chinese economy also implies the migration of younger and more skilled workers from rural to urban areas. Based on nationally representative rural survey data of 2002, covering 1,199 randomly selected households distributed across 60 villages in 6 provinces, it was found that most of the migrants from rural areas had been educated to, or above, junior high-school

Table 3.1 Demographics of long-distance long-term migrants in China

	Headcount	Income per month (CNY)	Months worked per year	Age	Education (years)
Migrants in off-farm employment beyond 6 months outside home county					
Total number	348	602	10.3	25.2	8.0
Monthly income over CNY 1,000	47	1339	9.9	28.6	9.1
Monthly income over CNY 800	85	1120	10.3	27.5	8.9
Migrants with off-farm employment beyond 6 months outside home province					
Total number	209	577	10.4	25.0	7.7
Monthly income over CNY 1,000	24	1266	10.1	28.1	9.0
Monthly income over CNY 800	45	1070	10.2	27.6	8.6

Source: Data collected by the Centre for Chinese Agricultural Policy, the Chinese Academy of Sciences, 2002.

level and were able to earn decent incomes in off-farm employment.[12] As shown in Table 3.1, in 2000 the average age of migrants working outside their home county for more than six months per year (defined as long-term migrants) was 25.2 years, with an average of eight years of education (a junior high-school certificate in China requires nine years). Among these long-term, long-distance migrants, 62.1 per cent were single and 63.2 per cent male. The average age and years of education of those with monthly incomes above CNY 1,000 was 28.6 and 9.1 years, respectively. Men dominated (87.2%) among high-income earners, 42.6 per cent of whom were single. This pattern was similar for workers earning more than CNY 800, and for those who moved outside their home county or province.

The relatively young age and good education of most rural migrants indicate a lack of both experience and interest in farming. It can therefore be expected that most of the young educated migrants would not value farming as much as the older and the less educated labourers left behind, and would tend to abandon the countryside for stable and decent urban employment and the chance to migrate to cities permanently.

Still using the 2002 rural labour survey data, we found that among 3,445 rural labourers surveyed, 1,581 were in off-farm employment, and 459 obtained off-farm employment outside their home county. As shown in Table 3.2, the labourers in off-farm employment outside their home county earned on average CNY 626 per month working an average of 8.6 months per year. Among the 459 labourers who worked outside their home county, 264 or 57.5 per cent, worked outside their home province, earning on average CNY 611 per month and working on average 8.9 months per year. Among the migrants with off-farm employment beyond six months, 348 had off-farm employment outside their home county and earned on average CNY 602 per month, working an average of 10.3 months per year. The 209 with off-farm employment outside their home province earned on average CNY 577 per month and worked an average of 10.4 months per year.

Table 3.2 Income distribution for long-distance long-term migrants

Total sample	Sample		National Income per Months						
	Number	% of total sample	Population (in mio)	month (CNY)	Worked (months)	Age (years)	Education (years)	Male (%)	Married (%)
	348	100	58.25	602	10.3	25.2	8.0	63.2	37.9
Income group									
Below CNY 200	32	9.2	5.36	112	10.1	23.1	8.7	56.3	21.9
CNY 200–400	61	17.5	10.21	290	10.6	22.7	7.0	62.3	21.3
CNY 400–600	105	30.2	17.58	479	10.5	24.7	7.6	58.1	32.4
CNY 600–800	65	18.7	10.88	655	9.9	26.4	7.9	63.1	47.7
CNY 800–1000	38	10.9	6.36	850	10.8	26.2	8.7	55.3	52.6
CNY 1000–1200	21	6.0	3.52	1046	9.8	26.7	8.7	81.0	52.4
CNY 1200–1400	13	3.7	2.18	1230	10.4	31.3	9.1	92.3	69.2
Above CNY 1400	13	3.7	2.18	1921	9.7	28.8	9.9	92.3	53.8

Source: CCAP Rural Labour Survey data and *China Statistical Yearbooks*.

Given that the sample is almost nationally representative, it is safe to infer that in 2000 there were 76.84 million rural labourers with off-farm employment outside their home county and 44.19 million outside their home province. If we limit ourselves to migrants with annual off-farm work for over six months, the corresponding numbers were 58.25 million and 34.99 million, respectively. These large numbers of long-distance migrants apparently cannot take care of their rural land. Furthermore, in the light of this group's relatively high earnings and stable off-farm jobs, they aspire to obtain an urban *hukou* and to gain access to urban social security.

The fact that China is still a country in economic transition also has profound implications for its policy options in *hukou* and social security reform. The *hukou* system was installed during the central planning period to minimize rural–urban migration. This resulted in a massive labour surplus in rural areas well into the late 1970s, when China initiated its economic transition. Even now, and despite increasing migration over the past two and a half decades, there is still a large surplus labour force in China's rural areas. According to the National Bureau of Statistics (2004), by 2003 the total labour force in China reached 760 million, of whom 256 million were working in the cities, 153 million in off-farm sectors in rural areas, and 325 million still active in agriculture. If the average GDP contribution per worker in non-agricultural activities is used as a benchmark, rural hidden unemployment is estimated at around 275 million (hidden unemployment being defined as low-productive work regardless of working time). If the benchmark is set more modestly at one-third of the productivity of non-agricultural workers (in line with other Asian countries), rural hidden unemployment would be around 150 million (Organisation for Economic Co-operation and Development, 2002). If we consider that around 5–6 million migrant farmers found off-farm employment in the past five years, it is reasonable to infer that China needs at least two to three decades to absorb its surplus rural labour force.

Given the large stock of surplus labour and the limited urban employment and infrastructure capacity, future reform of the *hukou* system needs to be implemented cautiously and incrementally. Although the *hukou* system has restricted China's permanent migration and aggravated its urban–rural disparity, it has indeed lessened the 'pull' force to urban centres and helped to avoid high unemployment and the spread of urban slums found in many other developing countries in South Asia and Latin America.

Economic transition also poses challenges for the creation of social insurance schemes for migrant workers. As a country in economic transition, China is struggling to implement a basic social insurance system in the cities and has yet to fully cover the country's urban population. More experience with social security systems is needed to successfully incorporate the migrant worker population into the existing urban social insurance schemes, or to design appropriate new schemes aimed at migrant workers in particular.

One good example is the pension system. The most serious challenge here is a serious funding shortfall largely attributable to the large-scale SOE restructuring in the second half of the 1990s. A chronic shortage of funds in the basic account has induced social security agencies to tap into personal accounts, originally intended as a reserve fund to deal with future aging problems.[13] Since the current system cannot guarantee future delivery of the promised benefits, it is very difficult to expand the pension system to the informal sectors (Zhao and Xu 2002). Therefore, it is crucial that China deals with the current difficulties accumulated in the process of economic transition before the country can effectively extend social insurance to migrant workers.

Inadequate reforms and the need for coordinated policies

Inadequate reforms so far

Since the mid-1990s, local reforms in the *hukou* system were carried out in small towns and some medium-sized cities. In a few large cities, such as Zhengzhou in Henan province and Shijiazhuang in Hebei province, the local government has begun to lower entry barriers for permanent migration. For instance, in Shijiazhuang, the capital of Hebei province, the city government adopted a policy in 2001 which granted a *hukou* to migrants who had been employed in the formal sector for one or two years, or who had technical qualifications, or who had a college diploma, or who had purchased local housing or who had invested in the city.[14] As of 2004, the central government also required local governments in migrant-receiving cities to ensure equal access to schools for the children of migrants.

However, in small cities the *hukou* reforms met with little enthusiasm from migrants because these cities were not attractive. At the same time, obtaining an urban *hukou* in most large and medium-sized cities and across provinces has remained difficult or impossible for most migrant workers. Thus, in most large and medium-sized cities, migrants need to purchase a property in the city and pay a large fee for the use of the urban infrastructure and facilities to be eligible to apply for a residence card.

The Chinese government is also initiating a pilot pension reform with the aim of achieving a transition from the pay-as-you-go system to a system with genuine individual accounts. The proposed changes would withhold 8 per cent of employees' salaries, down from the former 11 per cent, and channel these amounts into personalized pension accounts. The government is now considering setting up a special pension scheme for migrant workers characterized by purely individual accounts. Moreover, the new regulations would make it nearly impossible for the government to borrow money from personal accounts. This would be an important step forward, as government borrowing has practically gutted existing personal accounts.

However, so far government action in *hukou* reform has been unimpressive. Bold *hukou* reform policies mainly take place in smaller cities, where employment opportunities are limited. The few large cities that had opened their doors to migrant workers earlier soon found local infrastructure capacity and the provision of public services to be inadequate to meet the demands of incoming migrants, and had to lift their qualification criteria again for urban *hukou* applicants (Wang and Liu 2006).

Concerning social insurance schemes, especially the reform of pension schemes, current policies have not effectively addressed the issue of inter-regional transferability. Though many provinces are making efforts to expand the pooling of pensions from the municipal to provincial level, the technical difficulties and the possible fund redistributions among different localities have yielded very limited progress so far, not to mention the issue of (basic account) transferability across provinces.

Experimenting with reforms and transferring responsibility to the local level may not necessarily be a bad thing, given China's immense regional variety. Indeed, responses that address particular institutional situations and enable adaptation according to local circumstances would allow the necessary flexibility to respond to the particular hardships faced by different population segments in different places. For instance, concerning unemployment, some coastal regions are able to fund unemployment insurance for migrant workers, while in the interior provinces labourers generally have some land as a last resort. However, a localized and piecemeal response to *hukou* and social security reform relieves the centre of its responsibility, even though it is precisely the centre that has the necessary administrative advantage and capacity to coordinate action and reforms across localities through its policies and transfers, the mobilization of resources and pooling of risks necessary to effectively address the issue of social insurance coverage. To delegate responsibility to localities without also providing sufficient fiscal and administrative coordination means that vast numbers of people receive only superficial help or no help whatsoever (Murphy and Tao 2006).

Further reforms through coordinated policies

The Chinese government realizes that it is facing major challenges in its further reforms of the *hukou* system and the expansion of social security to migrant workers. These challenges arise from long-standing and deeply rooted structural imbalances that are not easy to resolve. It is necessary, therefore, to take these into account in an integrated policy reform, alongside flexible and particularized policy responses.

Further *hukou* reform must, therefore, resolve the issue of ensuring legitimate urban rights to an economically sustainable level to enable a reasonable degree of public services, including social assistance, housing subsidies and access to urban public schools to be extended to more migrant workers and their families. Given that an existing institutional framework regarding social services, subsidized accommodation and access to urban public schools is already in place in most Chinese cities, there is no compelling reason for the gradual expansion of such services to long-distance migrant workers to pose major administrative difficulties, provided the necessary financing can be secured.

Therefore, the key to reforming the *hukou* system is to establish an effective financing mechanism to enable municipal authorities in migrant-receiving regions to provide these services to migrants. As argued by Tao and Xu (2007), this would also make it possible for some migrating farmers to give up their land in the countryside, which would result in some extra land being released in migrant-sending regions and this in turn would enable the enlargement of existing farms to accommodate demographic changes in villages. Potentially, the government could collect some revenue on land appreciation in the process of land requisition, and use this revenue to fund social security schemes for migrant workers. Given that the value of agricultural land tends to increase significantly when land is converted to urban uses, and that at least part of the appreciation can be attributed to urban growth and infrastructure development, mobilizing some revenue from such land appreciation is justified.[15] Such a policy coordination mechanism would not only assist a breakthrough in *hukou* reform, but would also help to address the existing issue of land tenure insecurity in the countryside.

The Chinese government has made it clear that accelerating employment growth and extending social insurance schemes to cover a larger proportion of the population are at the top of its economic and political agenda. However, what is needed is effective and appropriate action to create and implement feasible and nationally coordinated policy packages that integrate the rural population into a wider system of social insurance regardless of their status. In light of the wide variety among migrant workers in terms of their employment, income and life plans, future reform policies should be directed to providing different scheme options for different groups of migrant workers. Migrant workers with relatively high and stable incomes in formal urban sectors should be encouraged to change their *hukou* status and to participate in the existing urban social insurance schemes. For migrant workers with relatively low incomes who may eventually return to the countryside, separate social insurance schemes might be designed to suit their particular needs.

With regard to work-related injury insurance, the present policy framework is relatively reasonable and the government should focus on policy enforcement in order to achieve a much higher coverage for migrant workers. Regarding unemployment and medical insurance, appropriate programmes that are specially designed for migrant workers might be a better choice. Such specialized programmes could, in principle, take the form of low-rate social pooling schemes funded by employers' contributions. For example, a medical insurance system funded through employers' contributions and covering only hospitalization costs for serious diseases would apply to most localities.

Given the advantages of resource pooling at higher levels, the responsibilities of providing work-related injury, medical and unemployment insurance would preferably be placed at higher government levels, such as the provincial or even the central government. This would require further fiscal reforms that shift social insurance responsibilities upwards to improve risk pooling and promote interregional equity. Such reforms would also push local governments at city level to focus more on social assistance, housing security and school services for migrant workers, thereby accelerate the *hukou* reform.

With regard to old-age benefits, a new pension system with consolidated individual accounts could be set up more easily if the issue of financing for SOE retirees' pensions were addressed, for example by selling the assets of the existing SOEs. This would facilitate the collection of fees from informal sectors and migrant workers and increase the coverage of current pension schemes. Along with progress in *hukou* reform, the migrant workers who obtain urban *hukou* status could then be incorporated into the existing pension system. At this stage, a special pension scheme with only individual accounts would be a better choice for temporary migrant workers. However, given the high funding shortfall in China's present pension system and the difficulties of collecting fees from the informal sector, any progress towards expanding pension scheme coverage to migrant workers can only be gradual.

Conclusions

China's economic reforms over the past two and a half decades have been described by its leaders as 'feeling for stones to cross the river'. However, the upheavals produced by various institutional changes have now reached a stage where this piecemeal approach is no longer sufficient. With incomplete transition, China is still highly segmented with obvious boundaries between urban and rural areas, and formal and informal sectors. Although past employment growth has been impressive, a sizable surplus of labour still exists in the rural sector. Increasing rural–urban migration, along with the restructuring of the urban sector, is generating huge pressure on employment growth.

With the still-functioning *hukou* system, most migrant workers cannot settle down permanently in cities and enjoy basic social services, housing security and public schooling for their children. They still suffer from separation from family members, and have to leave the old and the young in the countryside uncared for. If no breakthrough is achieved in *hukou* reform and the state policies continue to be biased towards urban and coastal areas, the rural–urban and coastal–inland disparities are set to increase further.

With further liberalization in urban labour markets, China has seen a gradual erosion of the traditional welfare system based on the work unit. After a period of piecemeal reforms and localized experiments, China has, over the past few years, endeavoured to establish and extend social insurance schemes, previously confined to state and urban collective sector employees, to all urban salaried workers regardless of the ownership status of the work unit. However, the success so far is very limited, both on account of the failure to enforce existing policies, and because these policies fail to meet the needs of migrant workers. Without adequate social insurance coverage, most migrant workers still face miserable conditions in the event of unemployment, work-related injury and illness.

Therefore, the main challenge for China in the coming years is the reform of the *hukou* system and to extend social security coverage so that the country's labour market can become increasingly integrated. This would not only help to generate more quality jobs for rural surplus labourers, but also offer better labour protection, social security and urban public services. To address these challenges, well-designed and correctly and promptly implemented policy packages are badly needed. Only through a holistic approach will more and more migrant workers from rural areas be able to settle in cities permanently and enjoy the basic rights to social security and urban public services.

Notes

1 The *Hukou* system in China is in effect an internal passport system. A person's local 'citizenship' and residence are initially determined as a birth right, traditionally determined by the mother's place of legal residence. Legal residence in a city entitles one to local access to permanent jobs, regular housing, public schooling, and public health care in that city. Until the early 1990s, it also entitled urban people to 'grain rations' – rations of essentials such as grain and kerosene. Legal residence in a village entitles residents to land for farming and residential land for housing, and access to local health and schooling facilities in rural areas. To permanently migrate to cities and be eligible for urban benefits, one had to change the legal residence status (see Chan (1994) for a detailed description).

2 The blue card functions more like the regular *hukou*; its holders enjoy most of the community-based benefits and rights. These individuals have the same local wages, resident tuition for elementary and middle schools, political rights, and most importantly, the chance to obtain a regular urban *hukou* in two to five years. To gain an blue-stamp *hukou*, migrants usually needed to pay a one-time entry fee – namely the urban infrastructural construction fee, which varies from a few thousand *yuan* in small cities and towns to CNY 50,000 in more attractive cities – they need to invested in a local business, or bought an expensive house. Between 1990 and 1994, local governments sold about 3 million urban *hukou* at an average price of CNY 8300 a piece (Liu 2005).

3 During this period, TVEs in rural areas also began to lose momentum in growth because of under-capitalization, an inability to spatially agglomerate as well as stronger competition from urban private sectors, foreign funded enterprises as well as imports, which also contributed a less dynamic rural growth (Brooks and Tao 2003).

4 A survey of the floating population in Shanghai found a clear division between them and local residents in terms of occupational composition, living conditions and income and benefits (Feng *et al*. 2002). According to a report in the *Beijing Daily* (10 April 1997), the Labour Bureau of one of Beijing's districts stipulated that at least 35 types of jobs should be closed to the floating population. Migrants without a local *hukou* are often expelled by urban authorities simply because they are outsiders and, therefore, considered to be potential elements of instability and crime.

5 In another study of occupational segregation and wage differentials between urban residents and rural migrants in Shanghai, Meng and Zhang (2001) found that rural migrants are treated differently from their urban counterparts in terms of occupational attainment and wages, after controlling for productivity-related characteristics, such as education, gender and work experience. They also found that around 22% of urban residents who would have been better suited for blue-collar jobs were given white-collar employment, while 6% of rural migrants who would have been suitable for white-collar jobs were relegated to blue-collar positions. As Meng and Zhang (2001: 487) noted: 'Not only do migrants take low-end jobs, but when they work in the same enterprise as do urban workers and perform the same kind of work, they also appear to be paid less.'

6   The lack of social insurance for employees is especially serious in TVEs. There is ample evidence that total reported earnings may capture almost all of their total compensation, because TVE workers do not have many of the social insurance and other welfare benefits of urban employees. At least by the end of 2002, the number of rural and small-town workers with any rural social pension insurance was minuscule. China's urban towns and rural areas have very weak or non-existent social benefit systems for pensions, medical insurance, unemployment insurance, workers' compensation and the like. Pension and medical insurance systems paid into by employers and employees are virtually non-existent in China outside of cities.

7   In Guangzhou and Dongguan city of Guangdong province, every year the city governments gain CNY 600 million and CNY 400 million, respectively, in the basic account because of the exit of migrant farmers from the pension system (Liu 2006).

8   Moreover, the current regulation requires employers to apply for termination of the social insurance account on behalf of the rural workers. Since employers have to handle the application for withdrawal of social insurance contributions, they are reluctant to incur these extra administrative costs. As a result, it is also likely that rural workers will leave their job without cashing in their contributions.

9   The national provisions regarding the time limit during which benefits are paid are as follows: where an unemployed person and the former employer have continually paid unemployment insurance premiums for more than one year, but less than five years, that person is eligible for benefits for up to 12 months; if the premiums have been paid for more than five years, but less than 10 years, the unemployed person is eligible for benefits for up to 18 months; if premiums have been paid for more than 10 years, the unemployed person is eligible for benefits for up to 24 months. Usually the level of monthly benefits is set at 70–80 per cent of the local minimum wage.

10 Industrial safety has been a long-standing problem in labour protection in China. Wonacott (2003) offers one of many reports that confirm a high incidence of accidents and workers' injuries in specific enterprises and localities. Between January and November 2003 alone, more than 120,000 people died in work-related accidents. In the first half of 2004, total work-related accidents reached 439,391, causing over 60,000 deaths and a daily death toll of 350 (National Bureau of Safe Production 2004). Safety concerns have figured prominently in the recent closure of large numbers of small-scale coalmines. In 2002, the State Council was in the process of drafting legislation on industrial safety for submission to the National People's Congress (Rawski 2002).

11 The Minimum Livelihood Guarantee (MLG) is one of China' 'three guarantees' system provided to SOE laid-off workers and low-income urban residents. Since 1998, the Chinese government has put into operation a system providing three guarantees: a guarantee for basic livelihood for laid-off persons from state-owned enterprises; guaranteed unemployment insurance, and guaranteed minimum living standards for urban residents. Laid-off workers receive a basic living allowance for up to three years. If they have not found a job by then, they can receive unemployment payments. If the per capita income of a family is below the local minimum living standard, they can apply for the minimum living standard guarantee for urban residents (Ministry of Labour and Social Security 2004).

12 The data are collected by the Centre for Chinese Agricultural Policy (CCAP) at the Chinese Academy of Sciences at Beijing in 2002. Using the same and earlier data, de Brauw *et al.* (2002) found that the age of migrants tended to decrease over time. In 1981, the off-farm labour participation rate (LPR) in rural areas for all age groups was within a small range of 18–19%. In 2000, the off-farm LPR for those aged 16–20 was as high as 75.8%, three times that of 1990 (23.7%). Off-farm LPR for the age groups of 21–25 and 26–30 doubled from 1990 to 2000. Off-farm LPR for those aged above 30 also increased from 20.6% to 37.6% between 1990 and 2000. For all rural migrants aged below 30, off-farm LPR increased from 31% to 45%.

13 According to the Ministry of Labour and Social Security (2006), as of 2004 some CNY 740 billion (US$ 92 billion) had been 'borrowed' from those accounts, and the deficit was growing by CNY 100 billion (US$ 12.4 billion) each year.
14 However, Shijiazhuang is exceptional as migrants only account for 5.07% of its total population, the lowest share in large cities in coastal regions, and lower than in most capital cities in inland provinces. Cities with a higher share of migrants in the local population still lag far behind in their reform (Li 2003).
15 As argued by Tao and Xu (2007), precisely because migrant workers also contribute significantly to economic growth in cities and thus also the land value appreciation in urban expansion, city governments should take some responsibilities to provide some basic public services such as social assistance and school education for migrants. As a matter of fact, a faster growing city tends to have faster urban expansion and higher land value appreciation, and thus would be able to collect more revenue from the proposed land value added tax. At the same time, faster economic growth tends to create more jobs and attract more migrants, thus the value added tax revenue and the number of migrants would tend to match each other and land value-added tax revenue can constitute the financial basis of the welfare package for new migrants.

References

Au, C.C. and Henderson, J.V. (2002) 'How migration restrictions limit agglomeration and productivity in China', NBER working paper no. 8707, January 2002.
Banister, J. (2005) 'Manufacturing earnings and compensation in China'. *Monthly Labor Review.* August: 22–40.
Beijing Government Research Group (BJRG) (2006) 'Local public service provisioning for migrant workers in Beijing' in *A Collection of Research Reports on China's Migrant Workers* (Zhongguo nongmingong diaoyan baogao). Beijing: Yanshi Press.
Brooks, R. and Tao, R. (2003) 'China's labor market performance and challenges, Occasional papers no. 03/210, International Monetary Fund, Washington DC.
Cai, F. (2000) *The Mobile Population Problem in China* (Zongguo liudong renkou wenti). Zhengzhou: Henan People's Publishing House.
Chan, K.W. (1994) *Cities With Invisible Walls.* Hong Kong: Oxford University Press.
Chan, K.W. and Zhang, L. (1999) 'The Hukou system and rural–urban migration in China: processes and changes', *The China Quarterly*, 160: 818–855.
de Brauw, A., Rozelle, S., Zhang, L., Huang, J. and Zhang, Y. (2002) 'The evolution of China's rural labour markets during the reforms: rapid, accelerating, transforming' *Journal of Comparative Economics*, 30(2): 329–353.
de Brauw, A., Huang, J., and Rozelle, S. (2004) 'The sequencing of reform policies in China's agricultural transition', *Economics of Transition*, 12(3): 427–465.
Feng, W., Zuo, X. and Ruan, D. (2002) 'Rural migrants in Shanghai: Living under the shadow of socialism', *International Migration Review*, 36(2): 520–545.
Fleisher B.M. and Yang, D.T. (2003) 'China's labor market' in Hope, N. (ed.) *Market Reforms in China*. Stanford: Stanford University Press.
Han, J. (2005) 'Unemployment and social security for land-losing farmers', working paper, State Council Development Research Center, Beijing.
Huang, P. and Pieke, F.N. (2003) 'China migration country study', paper presented at the Conference on Migration, Development and Pro-Poor Policy Choices in Asia, 22–24 June, Dhaka.
Kwong, J. (2004) 'Educating migrant children: Negotiations between the state and civil society' *The China Quarterly*, 180: 1073–1088.

Li, R. (2003) 'Distribution of migrants and *hukou* reform' (Wailai renkou fenbu yu huji zhidu gaige tantao), *Market and Population Analysis (Shichang Yu Renkou Fenxi)*, 4: 13–20.

Lin, J.Y. (1990) 'Collectivization and China's agricultural crisis in 1959–1961', *Journal of Political Economy*, 98: 1–10.

Lin, J.Y., Cai, F and Li, Z. (2003) *The China Miracle: Development Strategy and Economic Reform*. Hong Kong: Chinese University Press.

Liu, Z. (2005) 'Institution and inequality: the *hukou* system in China', *Journal of Comparative Economics*, 33(1): 133–157.

Liu, W. (2006) 'Assessment of some local social security experiments for migrant workers', in *A Collection of Research Reports on China's Migrant Workers* (Zhongguo nongmingong diaoyan baogao). Beijing: Yanshi Press.

Meng, X. and Zhang, J. (2001) 'The two-tier labor market in China – occupational segregation and wage differentials between urban and residents and rural migrants in Shanghai', *Journal of Comparative Economics*, 29(3): 485–504.

Ministry of Civil Affairs (2004) *Statistics Reports for Development of Civil Affairs in China*. Beijing: Ministry of Civil Affairs.

Ministry of Education (2006) 'Research report on education for migrants' children in China', in *A Collection of Research Reports on China's Migrant Workers*(Zhongguo nongmingong diaoyan baogao). Beijing: Yanshi Press.

Ministry of Labour and Social Security (MOLSS) (2006) 'Research report on social security for migrant workers in China', in *A Collection of Research Reports on China's Migrant Workers* (Zhongguo nongmingong diaoyan baogao). Beijing: Yanshi Press.

Murphy, R. and Tao, R. (2006) 'No wage and no land: new forms of unemployment in rural China', in Lee, G. and Warner, M. (eds) *Downsizing China: Unemployment and Market Reform*. London: Routledge, pp. 128–148.

Murphy, R. (2004) 'The impact of labor migration on the well-being and agency of rural Chinese women: Cultural and economic contexts and the life course' in Gateano, A.M. and Jacka, T. (eds) *On The Move: Women in Rural-to-Urban Migration in Contemporary China*. New York: Columbia University Press, pp. 243–278.

National Bureau of Safe Production (NBSP) (2004) 'Serious work safety situation in the first half of 2004', Xinhua News Agency, 20 July.

National Bureau of Statistics (NBS) (2002) 'Collection of data for China's fifth population census in 2000', *China Labour Statistical Yearbook*. Beijing: China Statistical Press.

National Bureau of Statistics (NBS) (2004) *China Statistical Yearbook 2004*. Beijing: China Statistical Press.

Organisation for Economic Co-operation and Development (2002) 'China in the world economy: the domestic policy challenges'. Organisation for Economics Co-operation and Development, Paris.

Pin, X. and Chen, M. (2004) 'Financing, land price and housing prices in Chinese cities', working paper no. C2004001, CCER, Peking University.

Putterman, L. (1992) *Continuity and Change in China's Rural Development: Collectives and Reform Eras in Perspectives*. Oxford University Press: New York.

Rawski, T. (2002) 'Recent developments in China's labour economy', report prepared for the International Labour Office, Geneva.

Solinger, D.J. (1999) 'Citizenship issues in China's internal migration: comparisons with Germany and Japan', *Political Science Quarterly*, 114(3): 455–478.

State Council Research Department (SCRG) (2006) 'Overview on China's migrant workers', in *A Collection of Research Reports on China's Migrant Workers* (Zhongguo nongmingong diaoyan baogao). Beijing: Yanshi Press.

Sui, X. (2005) *Investigations on Migrant Workers in China* (Zhongguo Minggong Diaocha). Beijing: Quyan Press.

Tao, R. and Xu, Z. (2007) 'Urbanization, rural land system and social security for migrants in China', *Journal of Development Studies*, 43(7): 1301–1320.

Wang, F. and Liu, W. (2006) 'A research report of several local hukou reform in several cities', in *A Collection of Research Reports on China's Migrant Workers* (Zhongguo nongmingong diaoyan Baogao). Beijing: Yanshi Press.

Wang, F. (2004) 'Reformed migration control and new targeted people – China's *hukou* system in the 2000s', *China Quarterly*, 177(March): 115–132.

Wonacott, P. (2003) 'Behind China's export boom, heated battle among factories', *Wall Street Journal*, 13 November.

Xinhua News Agency (2006) 'Medical insurance system covers 1.24 million migrant workers in South China'. Available at: http://english.people.com.cn/200602/26/eng20060226_246140.html

Yang, D.T. and Zhou, H. (1999) 'Rural–urban disparity and sectoral labor allocation in China', *Journal of Development Studies*, 35(3): 105–133.

Yang, Q. and Guo, F. (1996) 'Occupational attainment of rural to urban temporary economic migrants in China 1985–1990', *International Migration Review*, 30(3): 771–787.

Zhang, J. (2003) 'Urban *Xiagang*, unemployment and social support policies: A literature review of labor market policies in transitional China', report to the World Bank, Washington DC.

Zhao, Y. (2002) 'Earnings differentials between state and non-state enterprises in urban China', *Pacific Economic Review*, 7(1): 181–197.

Zhao, Y. and Xu, J. (2002) 'China's urban pension system: reforms and problems', *Cato Journal*, 21(3): 395–341.

4   Migrant children and migrant schooling

Policies, problems and possibilities

T.E. Woronov

The recent massive migration from China's rural areas to the cities has been the subject of much social and scholarly interest. Yet the millions of children who have migrated to the cities with their parents have received far less attention or research. Before 2001, when some of China's media, such as the influential nationwide *Nanfang Zhoumo* began to feature stories about these children, the vast majority of urban residents in cities like Beijing were largely unaware that very large numbers of children with rural *hukou*[1] were living in their city. Today, although the issue of migrant children and their education receives much more attention,[2] very little is actually known about them.

Government policies regarding migrant children and their education change very rapidly and vary by cities and provinces; for example, Beijing municipality and Jiangsu province issued new regulations about migrant children's education as recently as January 2006. To elucidate the effect of changing policies on migrant children, this chapter begins with an ethnographic description of an elementary school run by and for migrants in Beijing. Based on participant-observation research carried out in 2000–2001, the description of a migrant school, its students and teachers provides the context to understand and evaluate policies promulgated in 2006. The chapter concludes with a discussion of some of the most recent policy changes implemented in Beijing, and offers additional policy suggestions.

Migrant children

The phenomenon of large numbers of children living in cities without a residential *hukou* is very recent. Research (e.g. Davin 1999; Solinger 1999) indicates that the first large-scale movement of labourers from the countryside in the late 1980s and early 1990s consisted almost entirely of single adults; either men entering the service or construction industries, or women going into service or factory work. In either case, migrating individuals generally left their family members in their villages, with children cared for by remaining spouses or grandparents. But, beginning in the mid-1990s, as employment stabilized and rental housing became more widely accessible, increasing numbers of migrant labourers brought their families to join them in the cities.

Exact statistics of how many migrant children now live in China's urban areas are difficult to come by. In 2000, China's State Council Development Research Centre estimated that there were over 150,000 migrant children living in Beijing, of whom about 100,000 were of school age (Lu and Zhang 2004: 58), and these numbers have increased significantly since. The *Nanfang Ribao* (Southern Daily) estimated that in early 2005 some 1.5 million migrant children were living in Guangdong province,[3] while others claim that anywhere between two and six million school-age migrant children now live in cities across China (Kwong 2004).

These children are very diverse, coming from every province in China. Increasing numbers were actually born in Beijing and other host cities, yet are still considered to be 'rural' children. This reflects perhaps the one constant feature in the lives of all these children: migrant children are universally scorned by urbanites. Derided as dirty, dangerous and low quality (*suzhi*) (Douglas 1966), these children face serious social and cultural barriers that prevent their acceptance into urban life, even as official policies remove obstacles to their integration and education. The imagined and administrative dichotomy between peasants and urbanites is thus fundamental to how migrant labourers exist in the cities, and affects the daily lives of these children in urban areas.

Migrant schools

For financial reasons, children who either came with their parents from the countryside or who were born after their parents' arrival in Beijing, have been largely excluded from the capital's vast public education system. Before enrolling in a government school, a child without a residence permit (*hukou*) in Beijing is required to pay two sets of fees. One to the city of Beijing to obtain access to the system; in 2001 that fee was about 600 *yuan* (about US$ 75.00) per school year.[4] In addition, each family pays a separate fee (*zanzhu fei*) to the individual school or school sub-district to enrol their child. These fees vary widely from school to school, and can run from several hundred to several thousand *yuan* per semester. In addition, most schools also collect miscellaneous fees for extracurricular activities, books, supplies and materials. Not only are these fees prohibitively high for migrants, but many schools also require parents to pay several years of fees in advance. For migrant families who move frequently, these charges pose a major obstacle to obtaining school education for their children.

To circumvent these fees and offer education to children not legally registered in Beijing, migrants began to open their own private schools at rates affordable to the migrant population. There was only one such school in Beijing in 1993, but a decade later there were over 300 migrant schools in the capital, serving over 40,000 children (Wang 2003). Enrolment in these privately owned schools ranges from as few as 20 students to as many as 2,000, with fees averaging only 300 *yuan* per semester in 2001 (Han 2004: 36). A study conducted by the sociologist Han Jialing from the Beijing Academy of Social Sciences found that most of these schools served elementary school children; her survey of

Beijing migrant schools located only six junior-middle migrant schools, and only one senior-middle school.

Conditions in the migrant schools are generally very poor. According to a study conducted in 1998–1999 by the China Rural Labour Association,[5] although a few such schools rented space from standard government schools and thus had access to their facilities, the majority existed in much more tenuous circumstances. Researchers conducting the survey visited schools that rented space in car-repair shops, public bath houses, coal storage facilities or in the homes of migrant families.[6]

In the vast majority of cases, school facilities were rudimentary at best, with many not even having blackboards or chalk. As one researcher notes, owing to the poor quality of these schools, 'the migrant children's educational situation duplicates the unequal relationship between China's urban and rural areas, and undermines the principle of compulsory, equal and comprehensive education' (Lu and Zhang 2004: 79).

The legal status of these migrant schools has changed widely over time. While it is legal to operate privately owned schools outside the state sector (Chan and Mok 2001), completing the necessary transactions to be officially registered with government education authorities can be arduous and expensive, and many migrant schools are not formally registered. In order to better regulate these schools, a 1998 Provisional Decision by the Ministry of Education stipulated that local governments 'should assume some responsibility for migrants' children receiving the compulsory nine years of education', and should 'be relaxed' in their oversight of the conditions of such schools.[7]

However, by 2005 concerns about the low quality of migrant schools and the education they provided led several provinces and municipalities to lower barriers preventing migrant children from enrolling in local public schools. New regulations, discussed at the end of this chapter, reduced or eliminated fees so as to encourage more migrant families to send their children to local public schools. In order to contextualize and evaluate this push towards public education, the following section describes in detail the educational setting of one migrant school in Beijing.

Bright day school

In the winter of 1999, officials with the Ministry of Agriculture introduced me to one small migrant school in the Taiyanggong neighbourhood, north of the Third Ring Road in Beijing's Chaoyang district. Although located well within the city limits, until the early 1990s Taiyanggong was agricultural and very poor. Owing to the city's rapid expansion in the late 1980s and the legalization of rental properties, long-term residents of this neighbourhood were able to make more money renting out their tiny houses to newly arrived migrant workers than they could from tilling the soil. Eventually, most of the fields were replaced by lowly housing structures that local farmers rented out to sojourners from the rural hinterland.[8] In 2001, at least five migrant schools served the children of migrant labourers in Taiyanggong;

by late 2005, however, the neighbourhood's migrant housing had been demolished for redevelopment, and the migrant residents and their schools relocated north of the Fourth Ring Road.

The school I worked with, which I shall call the Bright Day School,[9] was owned and operated by a couple, Principal Chen and her husband, teacher Wang. Chen had been an elementary teacher and administrator in a school in rural Hubei province, in central China. As was increasingly common across this province, large numbers of able-bodied men, and then women, from her village had left for larger cities in search of work, leaving their children at home in the care of grandparents. As employment became more steady and their incomes rose, more and more of these labourers brought their children from their villages to live with them in Beijing. At one point, a representative was sent back to the village to ask the teacher Chen if she would also come to the capital to teach their children there. She agreed, recruited several young women from the village who were recent *zhongzhuan*[10] graduates, to go with her and her husband to Beijing and together they started a small school for children from their home village. After a few years, however, more and more children enrolled in her school because of its reputation for good quality teaching and reliability. By the time I arrived, the school had gone beyond teaching only students from one part of Hubei province and included children from all over China.

Upon her arrival, Chen's fellow villagers in Beijing arranged for her to rent a small courtyard in Taiyanggong. As was the case with most migrant schools in Beijing, Chen's new landlords facilitated connections between the newly arrived migrants and the local police in the neighbourhood. The Bright Day School was technically illegal, as it had never been registered with the local education authorities. Having opened her school without official approval – similar to most of the other migrant schools in the capital – Chen paid a regular fee to the local police to turn a blind eye to her operations,[11] typical of many such schools in Beijing.

The location of migrant schooling

In many ways, the Bright Day School was different from the regular Beijing public schools. The first and most obvious difference was the location of the school. Bright Day was extremely difficult to find, located at the end of a series of twisting, unpaved alleys north of the Third Ring Road. Unlike other schools and public buildings, nothing indicated or identified the school; a rusty metal gate at the entrance was locked from the inside with a padlocked bicycle chain.

The gate opened onto a small, walled area of approximately 300 square metres. Along the northwest side of the courtyard were two parallel rows of small dark rooms with south-facing entrances; these were the classrooms, plus one dormitory shared by the teachers and the school's few boarders. Chen and her husband lived in another small room against the northeast corner of the courtyard; it contained only a bed, a huge television and several cardboard boxes full of packaged snack food which she sold to the students during breaks between classes. Next to their room was a single water spigot and a small table on which another woman, Ms Liu, prepared three meals a day over an open coal fire (lunch for all the students and teachers; breakfast and dinner for the teachers, principal and boarding students). A dilapidated, netless ping pong table stood along the western wall, and an outhouse took up a section of the southeast corner. The rest of the small space was open for children to play between classes, at lunch and before and after school. Approximately 175 students and seven adults used this tiny space every day.

The Bright Day School ran from kindergarten up to fourth grade. When I first arrived there were six teachers, including Principal Chen, who taught the fourth grade and her husband, who taught kindergarten. There were two first-grade teachers, and one each for second and third grades. Only the third-grade teacher, a man, and Ms Liu, the cook, did not live on the premises. Chen was planning to add more teachers and classes for the next two school years, so that the currently enrolled students could continue through sixth grade.

The school's five classrooms were much more austere than those in Beijing's public schools. For example, the 23 fourth-grade students were packed into a dirt-floored 10 by 12 metres room; a single light bulb hung from the ceiling, with all other light coming in from the open door. The children sat on stools and used high, narrow benches as desks and a small, unevenly cut piece of slate on the front wall served as the blackboard. The room was so crowded that to observe Principal Chen when she taught this class, I had to perch on a stool outside the doorway. Yet this classroom was actually an improvement over those used by the kindergarten and first-grade classes, where close to 50 children were crammed into rooms approximately 15 by 17 metres large. The younger children sat on tiny stools, three children per each small table. The rooms were so crowded that the students in the front row were seated directly against the front wall, and had to look straight up to see the blackboard. The first-grade classrooms each had one small window which let in more light (and more wind), but all the other classrooms relied on whatever light came in through open doors. None of the classrooms had any heating in the winter or fresh air in the spring and summer.

Classes and schedules

These features were not the only differences between the Bright Day School and the state-run schools. Teachers at Bright Day taught all the lessons in the curriculum, unlike the standard Beijing public schools, where each subject was taught by a different teacher. And, unlike the standard Chinese schools, Bright Day teachers moved from one subject to another with great flexibility; no bells announced the end of one class period and the beginning of the next. Nor did Bright Day students follow a disciplined playtime schedule. Instead, teachers released their students to play in the open courtyard area next to the classrooms whenever there was a natural break in the teaching content, when students got too restless to sit still in class, or when a critical mass of other classes were let out (the noise from children from one or two classrooms playing immediately outside a classroom door was generally disruptive enough to cause the other teachers to declare recess as well). Unlike the regular Beijing schools, no formal or organized exercises were conducted during these breaks. Nor was there a set time for lunch; the children and teachers ate whenever Ms Liu had finished cooking. A distinct difference from public schools was, that rather than having to eat whatever was being served, the Bright Day students were offered a choice between the two or three tasty (mostly vegetarian) dishes Ms Liu prepared every day, and children who cleaned their plates were allowed to return for seconds. Thus, this school notably lacked many of the disciplinary elements characteristic of government-run schools (Foucault 1977).

School hours were also distinctly different at Bright Day. Unlike the standard schools, which unlocked the school gates at the legally mandated time of 7:45 a.m., at the migrant school students began arriving as early as 5:45 a.m. This was to accommodate the parents' work schedules. Many of the Bright Day parents worked in local vegetable markets that opened at 6:00 a.m., and needed to have their children supervised by that time. Nor were parents available to pick up their children at the usual school closing time of around 4:00 p.m., so that many of the younger children stayed at the school until 6:00 in the evening. This, I was told, explained the locks on the school gates: parents entrusted Principal Chen with their children for upwards of 12 hours a day, and she took seriously the responsibility of making sure none of them wandered off during that time. Their parents' work schedules also meant that the children rarely went to parks or other recreational spaces in the capital. Instead, these children worked most weekends: babysitting for younger siblings or nieces and nephews, cooking and cleaning, and helping out selling fruits and vegetables in the market.

Although the majority of children in this class came from Hubei, like Principal Chen, the other Bright Day students came from across the whole range of China's poorer, inland provinces, including Qinghai, Sichuan, Anhui, Guizhou and Inner Mongolia. All children's parents were employed in the 'secondary economy', which Zhang (2001b) defines as the range of 'private petty capitalist ventures and labour formations outside of state control', including restaurants, repair shops and construction. As noted above, many of the Bright Day students' parents worked in the local fruit and vegetable markets around the city, as well as in construction and catering. According to Chen, most were junior middle-school graduates, literate people who placed a high value on education. The parent's labour and educational history was typical of the majority of migrants in the capital (Lu and Zhang 2004: 66).

Bright day students

The Bright Day School students also looked quite different from their public school peers. One such difference was dress: Bright Day students had no uniforms and came to class in multiple layers of whatever they owned.[12] The children's clothes and faces were uniformly dirty, which Principal Chen explained by a lack of water. As was common in Beijing, none of the students' families had access to hot running water; but unlike most Beijing residents, the migrant families could only rarely afford the luxury of visiting the public bath houses. Even more striking than the lack of uniforms was that none of the Bright Day students wore red scarves, indicating their membership in the Communist Party-sponsored Little Red Pioneers. This lack was particularly noticeable because one boy in fourth grade took to wearing a red scarf, which he had bought himself. Chen told me that he bought these symbols of normative childhood so that he 'looked more like a regular (*putong*) student' in Beijing.

These students were also physically much more diverse, in particular because their ages were less consistent than among public school children. For example, the 23 students in the fourth-grade class ranged from age 10 to 15, and the third graders were aged 9 to 12; these age ranges were not uncommon among migrant children (Han 2004: 46). Principal Chen told me that many of the children in that class had travelled with their parents as they moved around the country looking for work, which caused them to miss several years of schooling. Others apparently had stopped attending school in their home villages after their parents had moved to the city, and this gap in their education set them behind their peers.

Gender distribution was another interesting facet of the Bright Day student body. In the higher grades, which enrolled fewer children, the gender ratio among the third and fourth-graders was almost equal. But in the kindergarten, first and second-grade classrooms, which had many more students, boys outnumbered girls by 65 to 35. I received several different explanations for this age-related gender difference. One official of the Ministry of Agriculture told me that this was merely incidental to the students enrolled at Bright Day, and that the gender ratio among children at other migrant schools in Beijing was more even. However, a teacher at Bright Day argued that this ratio was common: rural parents kept having children until they had a boy, so that there were many young boys with older sisters. The many boys that I saw in the lower grades all had older sisters who had been left in the home villages to care for themselves.[13]

I noticed another interesting difference between these children and the Beijing public school students. When I began working with the fourth-grade students to help them learn English, I noticed a very small child huddled on one of the students' benches. I learned that he was only five years old and was actually enrolled in the Bright Day's kindergarten class. He refused to leave his older sister's side, however, and thus attended fourth grade classes every day. This practice would have been inconceivable in a Beijing public school, not only because it violated the age-graded educational norms, but mainly because none of the Beijing *hukou*-holding children I ever met had any siblings. In contrast, among the older children at Bright Day School only one was an only child; all the others had siblings in Beijing or back in their home villages.

Teachers

Researchers Lu and Zhang (2004) found that most migrant families chose schools for their children based on location, preferring schools closer to their homes.

In contrast, Bright Day students and their parents frequently mentioned the quality of teaching as an important factor in their attending Bright Day School. In fact, several children travelled over one hour each way on foot or by bus each day to school. These children and their parents had formed close connections with Principal Chen and the teachers, and when the families moved to more distant parts of the city, they continued to send their children to Bright Day.

When I first arrived, there were six teachers at Bright Day, including Principal Chen and her husband. Three of the other teachers were young women in their early 20s, who were recent *zhongzhuan* graduates who had been recruited by Chen herself. These women all lived on the premises, sharing one tiny room with four students who boarded at the school during the week because their parents had moved to more distant parts of Beijing.

One of the teachers, Teacher Zhou, came to Beijing because she was unable to find work in her village. Teaching jobs there were highly coveted because they provided job security, but as fewer and fewer children stayed in the villages, teaching jobs were increasingly scarce. And, as a *zhongzhuan* graduate, she and her classmates were too highly educated to do farm work, but were not educated enough to do much else. As teachers in a migrant school in Beijing, they made very little money (about 300 *yuan* per month) and lived in crowded and difficult conditions, but had very high prestige in the migrant community as intellectual workers. Teacher Zhou and her colleagues told me that they enjoyed the 'opening to the world' they had experienced since arriving in Beijing, and the chance to learn more about urban life.

Observing their classes, I found that these young women were quite effective teachers who were obviously trained in the basics of elementary pedagogy, and who had formed close and warm relationships with their students. Several of their other colleagues, however, were less competent teachers. For example, Principal Chen's husband, Teacher Wang, had never taught before their move to Beijing; he controlled his kindergarten class largely by screaming at the children. The third grade teacher, Mr Wen, also had no teaching credentials, having been hired only because he was a senior middle-school graduate. He, too, had a more difficult time in classroom. Principal Chen was actively seeking replacements for both men, as well as additional teachers so as to expand the kindergarten and first-grade classes, but the low wages she paid made it difficult to hire and retain competent teachers.

Curriculum and lesson content

In spite of the structural differences in regimes of time and children's appearances, the Bright Day teachers all used the standard Beijing curriculum in their teaching. This was unusual among the migrant schools in Beijing, most of which imported textbooks from their home communities. This practice was designed to best prepare students for the test regimes in their native provinces, for it was assumed that they would complete their post-elementary education in the places where their *hukous* were based. Chen, however, was anxious to expand enrolment in her school, which she operated as a private, for-profit business, and felt that using the curriculum from her home province of Hubei would be off-putting for families from other parts of China. Not long after opening the school, she switched to the Beijing curriculum, apparently with the support of the parents.

The focus of daily lessons at Bright Day, however, differed from the Beijing public schools. For example, a significant amount of time in the lower grades was spent on drilling the students in standard Mandarin. As Chen explained, the children's rural accents were one cause for the derision they faced in Beijing, and Bright Day had an obligation to correct their pronunciation. For the same reason, Chen put a great deal of effort into teaching her fourth-graders to speak politely. She told me, 'Beijing children know how to use polite language, how to ask for things, how to say "May I borrow this?" instead of just grabbing things. These children's parents don't understand how to use polite language, so the students don't either. If they're going to live in Beijing, then they need to learn this well.'

Over the course of the year that I worked at Bright Day School, Principal Chen's interest grew in using the curriculum to make her students more closely resemble what she understood to be the norm among Beijing children. In late winter she reorganized the fourth-graders' class schedule to add one art class to practice drawing, and a class she called 'joyful education' (*yukuai jiaoyu*), her version of the increasing focus on 'education for quality' (*suzhi jiaoyu*) in the mainstream schools (Woronov 2003). Once a week, she gave the students a free period during which they were expected to 'create something', as a way to develop their creativity and thereby raise their 'quality'.

During this new class period children taught each other songs and dances from their home provinces, to write and perform skits, recite poetry, and draw pictures. Watching this class, I was extremely impressed by the students' creativity, their ability to observe and mimic the adults around them, and the wit they brought to their performances and poetry.

Mobility

In her survey of migrant schools in Beijing, Han Jialing noted that mobility was an important feature of all migrant schools (2004: 41). Because unregistered schools were subject to the whims of their landlords and the local police, they were frequently required to change locations. This was also the case at Bright Day School: their space was too small, but the landlord refused to allow them to add more rooms or expand into the adjoining buildings. By spring 2001, Chen had located a new space to rent and, one weekend in late May, the students and teachers moved the school two blocks north of its previous location.

The new school space was a great improvement. It was located in a large, relatively new enclosed compound (*dayuan*)[14] with a gated and guarded entrance that was hidden from the street. The compound had approximately 15 small rooms along the inside of each of the outer walls, and two interior rows of rooms in the centre with small alleyways running between them. Within the compound,

Bright Day School occupied a row of rooms around the southeast corner and along the south wall, including one room for Principal Chen and her husband, one for teachers and boarding students, and six classrooms. These classrooms were much larger than those in the previous location, and several had small windows which allowed in additional light. Across the small alley from the classrooms were the compound's indoor bathrooms with (cold) running water – an improvement many of the students commented on to me.

The other rooms in the compound were rented by families of migrant labourers from across China, who had different reactions to an elementary school moving into their midst. A few walked around with their hands clapped over their ears, scowling at the children's shouting and playing. But one family, whose room was located between the Bright Day classrooms, took the school's arrival as an opportunity for some entrepreneurship and began making and selling popcorn to the students during class breaks. Although the children were delighted, Principal Chen was not and she muttered that the smell of the popcorn was a distraction; perhaps because she resented the new competition for the snacks she sold the children.

Chen and her husband had used the opportunity of the move to expand the school's enrolment, and added a new teacher and a new kindergarten class. The following year she planned to add several more classes, to expand the first grade and add a fifth grade for the current fourth graders to enter. She also told me that she was contemplating opening a middle school as well, so that the students completing sixth grade in two years would have a place in Beijing to continue their education. However; middle-school teachers would be harder to come by, and she was also doubtful that the local authorities would give her permission to expand her operations to that extent.

The most interesting change in the move to the new school was symbolic: Chen and the teachers made considerable effort to make the new facilities resemble a standard school as much as possible. The small stools and benches were mostly replaced by desks. And, as in all standard schools, visitors to the new building were greeted by a large flagpole at the entrance flying an enormous Chinese flag. Chen told me that they had not yet begun to have the weekly flag-raising ceremonies mandated in all public schools (Landsberger 2001), but were planning to; in the meantime, she felt that the current flag looked 'official'. The classrooms did as well: each of the new rooms was neatly labelled, in careful calligraphy, with the grade level of the students inside – just as in the standard government schools. But the biggest change was in the students themselves. Suddenly, all the children in second grade and above were wearing Pioneer scarves, and several also sported new school uniforms. Chen told me that she was selling these to the students herself: she had located a school-supply store that sold her 'blank' polyester uniforms (with no school name imprinted on the back) which she resold to the children, along with the red scarves.

These improvements were not without risk, however. Chen told me over lunch one day that several others businesses wanted to expand into this new courtyard, and she anticipated that at some point the landlords would evict the school. To my surprise she did not sound angry or resentful, but instead was merely resigned. 'This is the way the system works', she told me, 'there's nothing we can do'.

Today, the Taiyanggong neighbourhood has been razed for development between the Third and Fourth Ring Roads. I have not been in contact with Principal Chen and the Bright Day School since the end of 2001, when the local police prohibited me from entering the school, so the fate of the Bright Day students, teachers and principal are unknown.

Advantages and disadvantages of migrant schools

In spite of the tenuous circumstances and the generally poor conditions in which they lived and studied, the students and teachers at Bright Day School were unanimous in telling me that they loved living in Beijing and loved attending the school. The reasons they offered were varied and interesting.

Most of the students told me that they greatly preferred living and studying in Beijing to the villages where they had come from. While their living conditions in the capital were more crowded and there was less space to play, they were together with their parents and, because of their parents' new urban incomes, the children ate more regularly and had better quality food. Several told me that although their parents had to work very hard, they were delighted that their families now had enough money to send them to school every day. The students also told me that even though the Beijing curriculum they studied at Bright Day was sometimes easier than the books they had used before, the quality of instruction at Bright Day was so much better that they were learning more there. One student said that the teachers in his home village spent all day playing *mahjong* rather than teaching; another noted that his former teachers did not read or correct homework, and that his learning time was better spent at Bright Day. Others reconfirmed that the combination of living with parents, who cared very much about their schoolwork, and their regular attendance with the good teachers at Bright Day, made them very happy.

The poor quality of education and care in the villages was a common theme among the migrant children, teachers and parents with whom I spoke. I was frequently told that the Bright Day children were getting a much better education in Beijing because they were living with their parents rather than under the ineffective watch of grandparents or other relatives. Village grandparents are not good caretakers, I was told, because as illiterates they do not understand the value of education, nor can they oversee homework. Others said that grandparents did not understand how to properly raise and control children today. Everyone agreed that having children live with their parents in the city was greatly preferable to leaving them at home. As one teacher said, she did not know what the future will hold for her Bright Day students but, whatever happens, they will have a tremendous advantage over those left in the villages.

Even though they were generally unable to avail themselves of the many cultural opportunities Beijing offered, the children very much enjoyed being part of life in a vast metropolis, and appreciated all of the life experience they had gathered as city dwellers. The teachers agreed, noting that the children's move to the capital had made them better students by stimulating their interest in their studies and their curiosity about the world around them, as well as their ambitions for the future. The Bright Day students, the teachers told me, were diligent, interested in their studies, and easy to teach; in spite of their difficult living conditions and low salary, the teachers also very much enjoyed their jobs and living in Beijing.

In my experience, these children were indeed much more sophisticated than their rural peers, and even many of their urban counterparts (cf. Lu and Zhang 2001: 60), particularly in their understanding and accepting of social difference. For example, rather than associating only with other children from their home villages or counties, the Bright Day students regularly told me that they particularly enjoyed school because it gave them a chance to meet and play with other children from all across China. I also found the children to be remarkably accepting of me, a foreigner. While many of the children in the Beijing public schools, where I also worked, were either frightened of me or refused to talk with me, since 'no foreigners can speak Chinese', the Bright Day pupils readily accepted me as just another outsider in the capital, one who also spoke heavily accented but comprehensible Mandarin, and who occasionally had strange but interesting ways of doing things. Others have noted this quality as well. Han (2004) reports interviewing a Beijing resident who chose to send her disabled child to a migrant school, because there the child was receiving attention and was well treated, whereas he had been neglected and discriminated against in the Beijing public schools. In fact, many of the Bright Day parents talked about discrimination and intolerance on the part of Beijing students and teachers. One parent, from Qinghai, told me that while she and her husband made enough money in Beijing to send their daughter to a regular public school, they had chosen Bright Day instead for fear that their bright and loquacious child would be looked down on by the Beijing children.

In addition, the atmosphere at Bright Day was much more relaxed and informal than in the regular Beijing schools I attended. The children seemed to genuinely enjoy being there and to like their studies, their classmates and teachers. The rigid hierarchies that segregated children of different ages and grades, and that kept children and teachers in different spheres, did not hold true at Bright Day. For example, at lunchtime the children ate wherever they wanted to rather than being required to sit at their assigned desk. Many times I would bring my rice bowl into an empty classroom at lunch, to be joined by a changing group of teachers, older students, younger children and sometimes Principal Chen. Together we might chat about life in the village, or have an impromptu English lesson, or the children would teach me songs from their villages. These kinds of informal interactions would not have been possible in the rigid structures of the public schools.

Yet, in spite of these advantages of attending a migrant school, the question of these children's future points to the many disadvantages they face. Although parents and children believed that Bright Day was providing a good education, migrant schools are by no means a panacea. One serious problem was the school's lack of resources. Unlike the Beijing public schools, Bright Day did not offer any non-academic classes – no gym classes, no formal art or music classes, and no computers. English was taught only when I joined the school. There were no extracurricular enrichment programmes that are so popular in the public schools, such as chess, calligraphy or martial arts, all of which are understood as essential in making children more well-rounded and higher 'quality' (*suzhi*) (Woronov 2003).

This lack of resources was especially poignant one day when a Bright Day fourth-grader arrived at school with a stack of children's magazines. At lunchtime, the children devoured these readings; nothing could tear them away. Later, they and their teachers told me that they all loved to read, but did not have access to any written material except their school books. The magazines – borrowed for one day from a student's neighbour – were the finest treat they could have hoped for.

Yet, beyond the lack of resources, the most serious problem faced by children in migrant schools is structural. For example, every migrant school uses a different curriculum, so that it can be very difficult for students to adjust if they change schools within Beijing or if they move to another city or return home. Much more serious, however, is the fact that the migrant schools are not registered with the local educational authorities (*jiaowei*), so that graduating students have no certificates to prove they completed their studies. These certificates are not merely an important social credential, for they – and the students' *hukou* – are also the only means for children to advance within the educational system. Even though it is technically possible for migrant parents to purchase places for their children in Beijing junior middle schools, without a *hukou* and middle-school graduation certificate, there is no mechanism for their children to take the city-wide *zhongkao* (senior high-school exam) to be placed in a senior middle school.

This is a real problem for migrant parents, and explains why there were so many more children enrolled in the lower grades at Bright Day than in the higher grades. Although Chen was eventually planning to add two more grades to accommodate fifth and sixth-graders, what would happen to students after completing those years was still unknown. Once their children finished second or third grade at Bright Day (or other migrant schools), many families sent their children back to their home villages to continue their education, reasoning that by age 11 or 12 they were old enough to care for themselves. Other children, however, remained in Beijing with their families after completing third or fourth grade, and began to work full-time as unskilled labourers.

Migrant parents therefore face a dilemma. On the one hand, they see their children's education as the best possible means of social mobility. But the only possible way to receive more advanced-level education is to return their children to the villages, where the education is second-rate and the children must live on their own. While many of the Bright Day children and their parents told me that their dream was to attend a university, the impossibility of even the brightest and hardest working students to reach that goal was surely a source of frustration. Since, as some researchers note, these children have a strong desire to excel in their studies as a way to pay back parents for their hard work and sacrifices (Lu and Zhang 2004: 61), the structural barriers that prevent them from advancing in their education must present serious emotional difficulties.

Psychological implications

Along with the structural constraints in their path, these children face an additional problem: they are despised in Beijing, at least among those adults who are aware of their presence in the city. People I spoke with in Beijing were unanimous in telling me that they did not want their children in the same classroom with migrants. Migrant children, I was informed, would inevitably be a negative influence on a school's 'learning atmosphere' (*xiaofeng*), and would only set bad examples for the urban children who were trying to learn. Most Beijingers I spoke with blamed the migrant children's parents, patiently explaining to me what they saw as patently obvious: rural people have low quality (*suzhi*), have little interest in or respect for education, and are unable to raise their children properly. My protests that the migrant parents and teachers I had met were deeply concerned about their children's education were summarily dismissed; every urbanite knew as common sense that rural migrants are dirty, unschooled and uncaring about education.

Noting this problem, several researchers have started looking specifically at the psychological aspects of migrant children's experiences in China's cities. Chinese Academy of Social Sciences sociologist, Wang Chunguang, has been interviewed on the experiences of what he calls the 'new urban generation' (cited in Shi 2004). His research found that second-generation migrants to the cities are not as willing to be transient as their parents were, and resist a future of moving from place to place with little constancy in their lives. Wang claims that the children he interviewed no longer fit back in their villages: they neither understand nor have any interest in farming, and even criticize rural people as being 'hicks' (*tai tu*) and 'unsanitary' (*bu weisheng*). Yet neither do they fit in the social world of China's cities, where they are discriminated against by local residents. As Wang states, if these children grow up not feeling that they belong anywhere, the result may be serious psychological problems for individual migrants, and a serious social problem for China as a whole. In a 2003 editorial, Wang argued that urbanization is not simply a process of making existing cities more modern, but also includes helping rural people to become urban (Wang 2003). This implies more than merely allowing rural migrants to live in cities to perform the dirty work that urbanites will not do, and instead requires that migrants and their children be made an integral part of the city. He says that providing education for migrant children will be an important means for this process to begin. If, instead, China's cities continue to forbid migrant children to enter city schools and ban the operation of migrant schools, the children only learn that urbanites neither like nor want them. What, he asks, will this mean in the future? How will this influence the future of Chinese society?

Others note this as well: Lu and Zhang (2004) point out that excluding migrant children from life in the cities, 'will bring further tensions into urban–rural relationships and solidify the disparities between the urban and rural areas. More seriously, it will cause migrant peasants and their progeny – a group of people with great vitality – to feel that they are being rejected and treated unfairly in the process of urbanization and modernization' (2004: 80).

Slowly, some urbanites are trying to achieve a better understanding between migrant children and resident Beijingers. In the fall of 2004, one organization affiliated with a local newspaper sponsored an activity called 'Holding Hands Together in Beijing' (*tong zai Beijing shou la shou*) which linked ten migrant children with ten Beijing families. Part of the activity was to grant each migrant child a wish. In a heart-wrenching story, the children told a reporter their wishes, which included: a visit the Forbidden City, trying a piece of birthday cake; seeing the tigers at the Beijing Zoo; owning a Chinese–English dictionary; getting a new book bag. The ten Beijing families fulfilled these children's wishes, showing how simple it may be to make these children more welcome in the city where they live and where many were in fact born (Shi 2004).

Policies and recommendations

Most Chinese researchers argue that increasing migrant children's enrolment in public schools will solve two of these problems. As more and more migrant children enter public schools and befriend local students, claims Han, the alienation and hostility they face in the cities will be reduced (Han 2004: 52). At the same time, the quality of their education will improve because they will no longer be restricted to studying in schools that are exempt from meeting local educational standards, or that exclude them from local and national testing for advancement and where they are subject to the profit-making interests of the school's owners (Lu and Zhang 2004). These researchers contend that municipal, provincial and central governments should work to reduce or abolish the fees that prevent most migrant children from accessing public education. As Lu and Zhang point out, state schools are responsible for educating children, but the fees collected by individual schools are not decreed by the state. As a result, 'in reality, these schools are using public goods to create private revenues' (Lu and Zhang 2004: 80).

In order to address this barrier to public education, in the past year several Chinese provinces and municipalities have forbidden public schools to charge any ancillary and special fees (*zanzhu fei*). Although these policies were originally intended to make education less of a financial burden for low-income urban families, in late 2005 and early 2006 several places – including Beijing – extended this policy to include migrant children. As reported in the *Jinghua Ribao* (Capital Times) on January 17, 2006, 'The policy of making (the nine years of) compulsory education totally free has been extended to children of migrant workers' in the capital. Other cities have used different methods to improve migrant children's access to public schools. The city of Nanjing, for example, has specified individual schools, located in or near neighbourhoods where migrant workers live where non-*hukou* holding children can enroll for free.

However, these new policies also have some serious limitations. One is that migrants who wish to enroll their child in a Beijing public school without paying any ancillary charges (*zanzhu fei*) must first present several different kinds of certificates, including a temporary residence certificate proving that they are living legally in Beijing, a document certifying that they have full-time work, and an education permit from the local education authorities (*jiaowei*). It is not known exactly how much these permits cost nor the kinds of administrative hoops that migrant workers must jump through to procure them.

In addition, the Beijing regulations only allow each family to register one child in school. While this is most likely not a burden for the vast majority of Beijing *hukou*-holders who have only one child per family, this stipulation will cause serious problems for rural migrants. Although there are no statistics currently available on how many rural families now have more than one child, I know that among the students I met at Bright Day School, only one was a single child. Presumably, families who want to take advantage of the new education policy will have to choose which of their children will be sent to school in Beijing. The effect this may have on girls' education – as well as that of younger sons – is still unknown.

At the moment, there does not appear to be any way to lower the barriers that these new policies create. The suggestion that urban schools could be open to any child who wished to enrol is deeply abhorrent to most urban Chinese, who believe that such a policy would lead to the total evacuation of the countryside, bringing all of China's peasants into the cities.

At the same time, even if migrant families can take advantage of the lower cost of public education, policies that allow more migrant children to enter the public schools do not address the problem of how these children may be treated once they are enrolled. In Beijing there are no provisions in the policy that mandate equal treatment for migrant children, nor is there any way to guarantee that individual classroom teachers will be held at all accountable for the ways they – and their students – treat migrants. And while Nanjing has designated five elementary and three junior middle schools to enrol qualifying migrant children without fees, it is still unclear how good these schools actually are, or whether they will be seen as a dumping ground for undesirable children.

As discussed above, some migrant parents claim that they chose migrant-owned schools for their children because of these problems in the public schools; others state that their children tried to study in public schools, but left because of the treatment they received. Migrant families thus face another dilemma in their educations decision-making: if their children study only among other migrants, they may grow up rootless, alienated and disconnected from the local people they live among. But if they do enrol in a public school, they may suffer discrimination and perhaps receive no better an education than that available in the migrant schools.

Perhaps one solution would be to find ways to encourage and improve the migrant-run schools (see Wang 2003). This may include requiring that they be overseen by local educational authorities in order to standardize curricula, testing and graduation requirements and certificates. Another solution to the problem of schools being operated for profit might be to turn over operations to non-profit organizations (perhaps along the model of some charter schools in the US), run through contributions by municipalities, migrant parents and charitable donations. This, too, would require tremendous oversight and careful monitoring; the question remains whether or not the finances and organization to carry out such an effort actually exist.

There is, then, no simple solution to the problem of migrant children's education. It is clear, however, that the opportunity to study is not a given for these children; they do not take for granted that they are in school, nor the standard life path to adulthood as understood by urban parents (Lu and Zhang 2004: 61). Although local and provincial governments are making efforts to improve the situation, it is also clear that millions of China's children are not receiving the nine years of compulsory education that the Chinese Ministry of Education has set as its goal. Neither is China fully meeting the Millennium Development Goal of providing basic elementary education for so many of its children, residents of some of the nation's wealthiest and most developed cities.

Acknowledgements

For their invaluable assistance in Beijing, I thank Lu Shaoqing, Zhao Shukai, Susan Champagne and the teachers, principal and students of the Bright Day School. Research on this topic was made possible through the generosity of the Committee on Scholarly Communication with China and the Spencer Foundation. I thank Norma Mendoza-Denton, Yang Fang and Fabio Lanza for their assistance and advice on this chapter.

Notes

1 The *hukou* is an internal passport system. A person's local citizenship and residence are initially determined at birth, traditionally by the mother's place of legal residence. To permanently migrate to cities and be eligible for urban benefits, one had to change the legal residence status.

2 An Internet search in December 2005 of the major Chinese-language media databases showed that by mid-2005, at least one article a day on the topic of migrant children (*liudong ertong*) was published somewhere in China.

3 *Nanfang Ribao*, 6 January, 2005, Internet document. Note that Lu and Zhang (2004) point out that migratory labour patterns are different in northern China than in the south; they therefore predict that more intact families with children will live in Beijing than in Guangdong.

4 *Nanfang Zhoumo* (Guangzhou), 7 June 2001, p. 1. (see also Han 2004). According to a Ford Foundation study done in April 1999, the average income of a migrant family in Beijing was 800–1200 RMB/month. Monthly expenses were almost equivalent so that some families had no money left at the end of the month, while others were able to save up to 300 RMB per month.

5 This is a division of the Ministry of Agriculture, which has some responsibility for rural *hukou* holders, including those living in cities.

6 'Survey of Schools at the City's Margins', Newsletter of Assistance of Migrant Children's Schools, China Rural Labor Association, 1 March 2000, p. 3.

7 *Nanfang Zhoumo* (Guangzhou), 7 June 2001; State Education Commission and Ministry of Public Security 2004 [1998].

8 See Liu and Liang (1997), Zhang (2001a) and Dutton (1998). These works discuss in detail the transformation of formerly rural parts of Beijing into migrant communities.

9 The names of the school, the students and teachers who worked there have all been changed to protect their anonymity.

10 *Zhongzhuan* is the common name for '*zhongdeng zhuanye xuexiao*' or three-year, technical senior high schools, particularly those responsible for preparing mid-level, white-collar workers. Students enter *zhongzhuan* upon completing ninth grade. Until recently, *zhongzhuan*-level teacher training schools supplied virtually all of China's elementary school teachers, although in Beijing and other large cities, more advanced qualifications are now required. In the rural areas, however, a *zhongzhuan* certificate is still considered a symbol of a high level of academic success (Thorgersen 1990).

11 See Zhang (2001b) for detailed descriptions of relations between migrants in Beijing and local police.

12 In Beijing everyone piles on layers and layers of clothes in the winter, so that the sometimes garish multi-coloured layered effect these children achieved was not in fact that unusual.

13 Han (2004: 48) speculates that the gender imbalance in the migrant school classrooms reflects a much broader gender imbalance in the rural population as a whole.

14 See Zhang (2001b) for a description of *dayuan* courtyard spaces for migrants.

References

Chan, D.K.K. and Ka-Ho, M. (2001) 'The resurgence of private education in post-Mao China: Problems and prospects' in Peterson, G., Hayhoe, R. and Lu, Y. (eds) *Education, Culture and Identity in Twentieth-Century China*. Ann Arbor: University of Michigan Press.

Davin, D. (1999) *Internal Migration in Contemporary China*. New York: St. Martin's Press.

Douglas, M. (1966) *Purity and Danger*. London: Routledge and Kegan Paul.

Dutton, M. (1998) *Streetlife China*. Cambridge: Cambridge University Press.

Foucault, M. (1977) *Discipline and Punish and The Birth of the Prison*. New York: Vintage Books.

Han, J. (2004) '[2001] Survey report on the state of compulsory education among migrant children in Beijing', *Chinese Education and Society*, 37(5): 29–55.

Jinghua Ribao (Capital Times) (2006) 'Dagong zidi yiwu jiaoyu yejiang xiangshou mianfei' (Migrant children will also enjoy free compulsory education), 17 January, p. 16a. Available at: http://edu.people.com.cn/GB/1053/4035059.html

Kwong, J. (2004) 'Guest Editor's Introduction', *Chinese Education and Society*, 37(5): 3–6.

Landsberger, S. (2001) 'Learning by what example?: Educational propaganda in twenty-first century China', *Critical Asian Studies*, 33(4): 541–571.

Liu, X. and Liang, W. (1997) 'Zhejiangcun: Social and spatial implications of informal urbanization on the periphery of Beijing', *Cities*, 14(2): 95–108.

Lu, S. and Zhang, S. (2004) '[2001] Urban/rural disparity and migrant children's education: an investigation into schools for children of transient workers in Beijing', *Chinese Education and Society*, 37(5): 56–83.

Nanfang Ribao (Southern Daily) (2005) 'Zhuanjia reyi wailai gongzinu jiuxue' (Experts debate the education of migrant children), 6 January.

Shi, X. (2004) 'Mingong di er dai' (The second generation of migrant workers), *Nanfang Zhoumo* (Southern Weekend), 2 December.

Solinger, D. (1999) *Contesting Citizenship in Urban China: Peasant Migrants, the State, and the Logic of the Market*. Berkeley: University of California Press.

State Education Commission and Ministry of Public Security (2004) [1998] Provisional regulations on schooling for migrant children and juveniles', *Chinese Education and Society*, 37(5): 7–9.

Thorgersen, S. (1990) *Secondary Education in China After Mao: Reform and Social Conflict.* Aarhus: Aarhus University Press.

Wang, C. (2003) 'Mingong zinü ruxue tixian shehui gongzheng' (Migrant children's education reflects society's justice), *Nanfang Zhoumo* (Southern Weekend), 10 April 2004.

Woronov, T.E (2003) 'Transforming the future: "Quality" children and the Chinese nation', PhD dissertation, University of Chicago.

Zhang, L. (2001a) 'Migration and privatization of space and power in late socialist China', *American Ethnologist*, 28(1): 179–205.

Zhang, L. (2001b) *Strangers in the City.* Stanford: Stanford University Press.

5 Migration in China

Reproductive and sexual health issues

Caroline Hoy

Introduction

A key difficulty faced by migrant workers in China is maintaining good reproductive and sexual health (Liang and Ma 1994; Zheng and Lian 2005). The problem is critical for at least three reasons. First, as the majority of migrants are aged between 16 and 35 years they are at a life stage when they are most likely to begin sexual relations, to engage in sex and to reproduce. Second, with socio-economic liberalisation, the rise of sex markets, contaminated blood scares and population mobility, the health risks from sexually transmitted infections (STIs) and HIV/AIDS are on the increase. Finally, migrants are more deprived than other urban social groups with regard to opportunities for gaining access to public health information and to affordable health services

The lack of reproductive health services for migrants in China is hampered by general assumptions about migrants' identities. There is a lack of understanding about the social worlds of migrants, and a failure to recognise the heterogeneity and the varying needs of the individuals who fall under the category of 'migrant'. Migrants tend to be seen only as transients in the cities who will eventually return to rural areas. Migrants also tend to be viewed primarily in economic terms – as labourers and as people who make claims on urban infrastructure – a perspective which overlooks that they are complete human beings with complex aspects to their lives, including sexuality. Any recognition of the sexual aspect of migrants tends to be, for the most part, in relation to migrant women who are seen as reproducers within the context of marriage; and here the focus is very much on preventing them from having unplanned births. This kind of perspective underpins a neglect of the role of migrant men in sexual relations and a neglect of the needs of all migrants who engage in sex outside marriage (Jing 2004).

This chapter explores the reproductive and sexual health needs currently facing migrants in China and reflects on why current provision for them continues to be inadequate. The literature on international migration and sexual health provides the introduction to this discussion because it has key lessons for domestic migration in China. Insights from this literature also underscore the need to acknowledge the existence of migrants' sexuality if their social development needs are to be adequately addressed. Parts two and three examine sexual behaviour in

China firstly in the general population and then with specific reference to migrants. These sections challenge the received wisdom in China that state migration and the transmission of STIs are inevitably linked. The following two parts contrast the focus on married women in government policy and practice with the realities of accessing information and services. The final parts examine the work of non-government organisations (NGOs) which can be both innovative as well as more traditional in approach, but which often retains a moralistic base. Good practice, however, requires that the vulnerabilities of migrants are recognised in pragmatic activities. The reform of reproductive and sexual health programmes and greater leadership are identified as crucial to the continued development of sexual health and information services in China. The chapter concludes by looking at these programmes and services in the context of the millennium development goals (MDGs). It is argued that by implementing programmes aimed at improving migrants' entitlements, China will ensure that the MDGs become more realistic aims for the whole of society.

Migration and sexuality: the international dimension

Work on migration and sexuality in the wider migration studies literature has been dominated by attention to migrants who move across international borders. Such work has examined the reactions of international migrants to different social contexts, and changes in their sexual attitudes and behaviour in response to new conditions. For example, a study of Iranians moving to Sweden has shown that single adult migrants were influenced by Swedish attitudes towards sexuality and relationships (Ahmadi 2003). This is instructive for researching migration patterns in China, because migration from rural to urban areas within China similarly involves moving over long distances and exposure to different traditions and cultural settings, as well as the emotional stress of separation from families and communities (Lou *et al.* 2004).

Where young women are part of migrant families it is often necessary for them to adapt to, and satisfy, the expectations of both their traditional family and their new host societies. This can be the cause of strains within the traditional family. For instance, it has been observed that daughters of Moroccan migrants living in the Netherlands were subject to stricter rules of conduct than young girls in Morocco (Buitelaar 2002). In Shanghai, young migrant women living with their families in the city were similarly subject to certain traditional family restrictions on sexual behaviour (Hoy 2004a).

Data from the Demographic and Health Survey (DHS) from countries around the world show that migrants are consistently viewed negatively by local populations and their sexual behaviour is generally judged by the host population to be more liberal and less moral than the norm. Yet, the DHS data also show that migrants are often less experienced and more vulnerable to sexual risks; they are often less aware than the resident urban population of available methods of contraception and protection against sexually transmitted diseases (STDs) (Gardner and Blackburn 1996). Specific country case studies of internal migration produce similar findings.

For instance, a study of migrants to free trade zones in Sri Lanka found that they were unprepared for the new and freer forms of sexuality in the destination area (Harcourt, 1997a; Hewamanne and Müller 1997).

Such experiences notwithstanding, care should be taken not to stereotype migrants as simple and unsophisticated compared to populations which have greater exposure to modern trends and social attitudes. Indeed, while they are unfamiliar with new social norms, they have their own varied and complex behavioural norms and traditions. For instance, a study of migrants in Nepal revealed that their attitude to risk, pleasure and sexual relationships predated their migration to Kathmandu. These young men and women had casual sex and engaged in risky behaviour while still in their local communities, despite the general disapproval of pre-marital sex in the local society. Opportunities in Kathmandu were simply more overt (Puri and Buszna 2004).

However, individual responses to new or different experiences and insights brought about through migration are conditioned by the personal profile of the migrant as much as by wider context (Ahmadi 2003; Wolfers *et al.* 2002). As individuals develop a migrant identity they also accept or reject other sexual behaviours, which may in some instances lead to apparently contradictory conduct, such as the denial of the existence of a spouse at home to overcome loneliness in the new location (Wolfers *et al.* 2002). By ignoring the fact of their wives' existence, migrant men can find company while avoiding the personal distress of consciously betraying their spouse. Social factors can increase the health vulnerability of migrant populations to the consequence of such new sexual behaviours. Examples of these social factors include taboos against openly discussing sexuality; ignorance about sexually transmitted infections and HIV/AIDS; and limited access to public health information and services (Johns Hopkins University Center for Communication Programs (JHUCCP) 1996).

Yet while research rightly explores the effect of disorientation and separation on migrant communities and their social conduct, it is also true that migrants have access to various forms of contact, support and advice, particularly within the migrant community. This social support is often rooted in home-based relationships that originally facilitated migration; the friends or family members who gave initial information about city jobs or other opportunities (Hsu and du Guerny 2000). However these support networks may exert contradictory effects when it comes to reproduction and sexual health; for instance, they may provide information about contraception matters on the one hand and may constrain open discussion of new sexual behaviours on the other.

The international migration literature suggests that programmes to provide help, information and support for migrants must come to terms with the reality of migrants' sexual behaviour and relationships and acknowledge the issue of the sexually active migrant (Wolfers *et al.* 2002). Besides offering assistance, such programmes should enable migrants to make informed choices about their options and the consequences of their choices. Programmes must also be responsive to the complex relationships migrants have with both their home and host communities. Moreover, programme and projects intended to assist migrants have to be aware of the existence of ideological bias and agendas embedded within the programme itself (JHUCCP 1996). Practical barriers that inhibit migrants from accessing services need to be identified. These include the longer distances migrants often have to cover and the resources required to obtain access to healthcare. Therefore, such programmes should aim at conveying the necessary information and aid to migrants, rather than leaving it up to the migrants' own initiatives and resources to request access to assistance (JHUCCP 1996).

While this may seem obvious, for many countries to acknowledge the existence of migrants' sexual needs requires a shift in the way the whole population itself is viewed – from being seen as economic labourers and social transients, to being seen as full members of the national community with complex and wide-ranging human needs. Reproductive health and its associated services are rarely mentioned in documents on general health reform in China let alone with specific regard to the needs of migrants.

Sexual behaviour in China

In China, particularly in urban areas, more liberal attitudes towards sexual behaviour started to become apparent and impact on society over the course of the 1980s (Hoy 2001a). One of the main drivers of change was the birth planning policy launched at the end of the 1970s. The birth planning policy severed the traditional relationship between marriage, the onset of sexual behaviour and childbirth (Hoy 2001a; Jefferies 2004; Zha and Geng 1992). However, it also excluded sexually active but unmarried individuals from receiving health and education services which were solely aimed at married couples (Friedman 2000; Ru 2005).

The changes in sexual behaviour in China have received much attention as researchers have investigated the speed of the transformation and compared sexual attitudes in the pre- and post-Mao eras. Such research activities are rendered more complicated by the fact that sexual attitudes, and practices and the nature and pace of change taking place in the population, differ across China and are subject to local conditions (Farquhar 2002). Nor is the liberal behaviour and overt displays by any one social group indicative of such changes and tangible cultural shifts occurring throughout the population as a whole. Alongside sexual practices, the related emotions and sentiments are changing too (Farquhar 2002). For instance, the Chinese have embraced western ideals of romantic relationships (Farrer and Sun 2003). Yet, such examples should not lead to the conclusion that contemporary Chinese sexuality is simply becoming more 'westernised'; rather, established behavioural and sexual norms are evolving within their own particular Chinese context (Farquhar 2002; Jefferies 2004).

The contrast between the more liberal attitudes and behaviours of the urban population and the more conservative attitudes and practices in rural areas is acknowledged; yet rural populations as a whole are not necessarily subject to the same conservative sexual mores (Friedman 2000), and sexual behaviour is likely to be more diverse and pragmatic than may be presumed. This has been shown in work based in rural Fujian, where the links between marriage and childbirth are very strong, but pre-marital sex and cohabitation are also tolerated (Friedman 2000).

Migrants and sexual behaviour in China

Migration and sexuality in China are often considered together in relation to what are perceived to be migrants' looser sexual attitudes and immodest behaviour. Notwithstanding changing moral attitudes in the general population, as suggested by a rising prevalence in pre-marital sex and prostitution (Evans 1997; Macmillan 2004; Parish 2002), migrants in particular are derided for their supposed questionable sexual morality. This is because migrants are thought to relinquish their traditional social and moral precepts when leaving their villages. And in the absence of stable moral bearings in their new and anonymous urban setting, they are more likely to engage in risky and wanton conduct (H. Yang *et al.* 2005a, 2005b).

Chinese academic research on the sexual behaviour of urban migrants has shown that sexually experienced female migrants working in the entertainment or personal service industries are twice as likely to engage in unsafe sex as migrant women working in the non-entertainment sector (H. Yang *et al.*, 2005a). Women active in the sex industry are typically young and unmarried, consume alcohol and drugs, and seldom use condoms (H. Yang *et al.* 2005b). What such studies do not show, however, are the circumstances and pressures that lead migrants into sexually exploitative work. Often, in the face of unemployment and a lack of support from either the state or family and friends, the sex industry may be the most feasible means of survival.

Another common theme associated with migration and sexuality in both the Chinese and Western literature on China concerns the increase of sexually transmitted infections, including HIV and AIDS (Parish *et al.* 2003; H. Yang *et al.* 2005a, 2005b; X. Yang 2005; Zhang and Li 1999). Migrants are reported to have low levels of knowledge about STIs (F. Wang *et al.* 2005). Though China had succeeded in eradicating STDs during the Mao era (Bannister 1987), many observers worry that the less inhibited behaviour of migrants may also be associated with an increase in the prevalence of HIV, STDs, intravenous drug use (IDU) and commercial sex (Wang and Gao 2000; X. Yang 2004, 2005) to the extent that a third HIV/AIDS epidemic in China is occurring through sexual transmission from high-risk groups such as IDUs, female sex workers, their clients, and the highly mobile internal economic migrants to the general population (Thompson 2004).

Moreover, there are concerns, that on account of the regular returns by migrants to their home villages, STI transmission to the wider population is likely to occur in both rural as well as urban areas. In fact it is thought that risks may be higher in rural areas because spouses in the village may be less likely to know about STIs or to question the fidelity or safety of sex with their partners. Murphy (2004) provides some rare evidence about this phenomenon, based on information provided to her by doctors in Jiangxi on the level of STDs suffered by migrants and their wives. She argues that this may be a particular problem where there is marked out-migration of men to urban factories, and less so where women accompany their husbands to participate in small enterprises. The impact of STDs is both physical and emotional in nature (Murphy 2004).

Although migration and the transmission of STIs are often linked in the literature and there are undoubted connections between population mobility and disease transmission, the casual associations remain complex (Xiang 2004). Hence, unqualified generalisations about migrant behaviour and moral attitudes must be treated with caution. Scholars have noted that STDs are prevalent in places where migrants tend to congregate because these locations are cheaper and less hostile to migrants. As a result, migrants may be numerous in places of high HIV/STI infection without being the cause or initiators of casual sexual conduct (Smith 2005). In fact, officials and businessmen in both rural and urban areas have been shown to be more likely than migrants to buy sexual services (Kaufmann and Jing 2002). Further, this concept of 'core transmitters' of HIV/AIDs, such as prostitutes and migrant workers has been recently shown to have damaged our understanding of the spread of HIV/AIDS in Africa. Recent work shows that while mobility may contribute to the spread of infection, the social context must be examined more thoroughly before migrants/prostitutes are further stigmatised in such a way that admission of HIV status, associated as it is with migration/sex work, becomes more difficult for the population more generally. A more thorough investigation of sexual behaviour may, as in the case of Africa, offend conventional wisdom, but lead to a more nuanced appreciation of social and sexual relationships (Easterly 2007).

There are two further reasons for using caution when examining the link between migration and infection. Both relate to forms of monitoring over migrants. Many migrants commonly have their behaviour scrutinised by fellow migrants from their hometowns and by family in the village (X. Yang 2000). Although it may have been the case in the 1980s and 1990s that a migrant's separation from family and the local community entailed a greater degree of isolation, in the 2000s the widespread availability of mobile phones has made it possible for many to maintain contact with family and friends at home. At the same time, as a result of the increasing prevalence of phones, it is more likely that gossip from migrant networks in the cities will reach family in the village. Migrants also commonly have their behaviour controlled through disciplined and gruelling labour conditions. Strong surveillance and strict work regimes as documented by Pun Ngai in this volume mean that there are few opportunities for male and female sexual interaction. Dormitories supplied by factory bosses are strictly separated (Xu 2000: 171) with one of the main objectives being to avoid pregnancies among migrant women workers and to maintain the economic efficiency of the workplace.

Chinese government approach to reproductive and sexual health for migrants

The government emphasises that it has been making every effort to provide healthcare services to migrant women (White Paper 2005). These efforts have included expanding the availability of services, such as medical checkups and access to contraception for migrant women in their destination communities, instead of making them return to their home villages for information and care. The paper also identifies HIV/AIDS as an area of concern to be earmarked for additional funding. The work of NGOs in this area of programming is acknowledged (White Paper 2005). The White Paper is built on the 'Outline for the Development of Chinese Women (2001–2010)', which listed a number of goals and policies to empower women in such areas as the economy, decision making, law and health. Yet the paper also emphasised that the devolution of strategies and plans has been enacted throughout the public system (White Paper 2005), a relatively under-funded and lacklustre apparatus with poor administrative capacity (Greenhalgh and Winkler 2005), suggesting that the challenges of translating policy into practice remain immense. Significantly, the role of men is mentioned only incidentally.

Yet even if the state is now paying more attention to needs of migrant women, as regulations in Box 5.1 show, the focus on married women remains deeply entrenched. And it is precisely this feature of state programmes – the persistent view that family planning services are not meant for the unmarried, widowed and divorced – which hinders the provision of reproductive and sexual health services to all migrants (Popinfo 2003). The ongoing focus on married women occurs in part because for so long, fertility regulation rather than sexual health protection and sexual lives has been the focus of service provision. The resulting exclusion of men and of young unmarried people from reproductive programmes has particularly serious consequences in China because the government is the main source of sexual health information (Hoy 2001a). At the same time, the unmarried migrant population is known to be sexually active, too embarrassed to seek family planning advice and further put off by the high cost of services (Chen and Zheng 2001; Hoy 2007; Liu *et al.* 2004; Wang and Gao 2000; B. Wang *et al.* 2005). The preoccupation of population planners with married women also means that despite the policy statements which require men and women to take equal responsibility for reproduction and contraception (such as that issued in 2003, see Box 5.1), these good intentions are seldom translated into practice. It will clearly take more concerted efforts and lobbying by feminist groups and NGOs before the sexual health rights and responsibilities of migrant men and women will receive sufficient attention and become integrated into legislation or education programmes (Berer 1998).

Sexual and reproductive health education for migrants in China

Young men and women make up a significant percentage of the world's population, and this is also true in China. Concern about reproductive health issues and the needs of adolescents was highlighted at the International Conference on Population in Mexico in 1984 (United National Economic and Social Council 2002). Given that young adults account for a large share of the world's population, many

---

Box 5.1 Example regulations on population and family planning

1   State Council Measures on the Administration of Family Planning for the Floating Population, effective January 1, 1999: [...] apply to currently married people of child bearing age.

2   People's Republic of China, Law on Population and Birth Planning, 2002, Chapter 6, Supplementary Provisions, Article 45:
    The State Council shall formulate specific guidelines for managing the family planning programme among migrants, specific measures for managing family planning services, and measuring the for the administration of the collection of social compensation fees.

3   State Council document on offering better management and services for peasants who migrated to work in urban areas. State Council document no. 1 (Policy Statement), 2005, reproduced in Huang and Zhan (2005).
    Priority will be given to the enhancement of the proper management of migrant populations, including family planning, education for children, employment, health care and legal aid.

4   Provincial Government of Heilongjiang, Order No. 8, method for the management of the floating population in Heilongjiang province, June 2, 2000:
    Article 10: Family planning publicity and educational work among the floating population will be the responsibility of the home and host cities.
    Articles 11 and 12: Married migrant women of childbearing age shall receive medical checkups and services concerning contraception, birth control and reproductive health related to family planning.
    http://www.unescap.org/esid/psis/population/database/poplaws/law_china/ch_record054.htm

---

international health activists have argued that it is important to understand the issues and needs of adolescents in relation to their entry into reproductive life and how they experience their sexuality. Low educational attainment and lack of skills prior to the first sexual encounter have been recognised as global problems, and the importance of education in relation to individual sexual initiation has been acknowledged (United National Economic and Social Council 2002). Generally, higher levels of education are known to be positively correlated with later sexual initiation, better information and a higher sense of responsibility because such young adults tend to have the necessary self-confidence to control their sexuality and avoid casual sexual encounters.

While there is general agreement in the literature that greater numbers of young people in China are engaging in sexual behaviour, that the age of sexual initiation is declining and that there is an unmet need for reproductive health education (UNICEF 2007), there is little information available on the impacts of education on age at first sexual initiation, especially where the migrant population is concerned. This lack of information is associated with the challenges of researching the sexual behaviour of young people. The small body of existing studies suggest contradictory findings. In one study, higher education was associated with STD diagnosis for migrants (Liu *et al*. 2005) while in another study low educational achievements were reported for migrants visiting STD clinics in three cities (Wang *et al*. 2007). A further finding (Zheng and Lian 2005) is that migrant women are more likely to have abortions at younger ages compared with the resident population and that these abortions occur in the later stages of pregnancy. Finally, a study suggests that specific sexual health education programmes have not necessarily been associated with a later age at first sexual intercourse (B. Wang *et al*. 2005).

While the family does play an important role in this period in young people's lives, information and advice about reproductive health and sexuality are not generally passed on from parent to child. In both developed and developing countries, young men and women are often left to their own devices to glean information from other people or through literature and the media (Buston and Wight 2006; Harcourt 1997a; Hewamanne and Müller 1997; Hoy 2007; Pitanguy and de Mello e Souza 1997). Often, learning about sexuality entails a coming of age through trial and error with all attendant risks and difficulties, as pre-marital sex among youth generally, and also young Asian women, is on the rise. In the case of rural China, it is also worth noting that further obstacles to the transmission of information about bodily care and processes from mothers to daughters and from parents to children more generally are presented by the long-term absence of migrant parents from their children who remain in the villages (Tsukamoto 2005).

In China, reproductive and sexual health programmes included in school curricula are generally problematic with respect to both their content and their presentation to young people, and it is even less likely that people schooled in rural areas will have received adequate sexual education (Hoy 2001a). Early attempts to introduce reproductive and sexual health education have been largely provisional and limited to certain schools or age groups. Comprehensive guidelines on the teaching of such subjects were not issued until 1997, and nor are they expected to be universally implemented. Neither China's 2000 Population and Family Planning Law nor the 2001 White Paper on the Population and Development of China in the 21st Century have referred to the provision of sexual and reproductive health education (Ru 2005). In 2003, addressing the issue of adolescent sexual and reproductive health, Zhang Weiqing, Minister of the State Population and Family Planning Commission emphasised the need to provide sex education, but also ethics to young people, with particular attention to be paid to sexual and reproductive health. This is stated in the Population and Family Planning Law (People's Republic of China Law on Population and Birth Planning 2002).

However, the information provided by parents and schools does not reach enough young people or meet their needs (Liu 1997). A research project from the 1990s found that among the targeted girls, two-thirds had received no sexual and reproductive health education prior to experiencing their first period. Where information was available it was only from very basic textbooks or magazines (Liu 1997), and more recent research in Shanghai confirmed that this was still the case. Fieldwork conducted in 2004 by the author found that only a few young migrant women considered the reproductive and sexual health education received at school to be useful. However, rather more young women reported even less helpful experiences; for example, they were made to read the textbooks on their own without any adult guidance or answers to any questions that they may have had (quoted in Hoy 2006: 8–9). Nor was there any consideration of different individual needs (Macmillan 2004; see also Wang and Van de Ven 2003).

On the other hand, where education projects were conducted in a sympathetic manner, they were welcomed by the participants and positively influenced their behaviour. One community-based sex education and reproductive health service programme in Shanghai led to increased use of contraception and protection among its young unmarried population (B. Lou *et al.* 2004). However, where generation gaps and incompatibility, or conflict among information providers prevent the dissemination of appropriate information and guidance, it is young, unmarried women who suffer most (Zhang and Locke 2002).

In China it has generally been women who have accessed formal sexual and reproductive health education – indeed most formal interventions target them. Men for their part have tended to use informal sources of information, which are often inaccurate and misleading. Young Chinese men living in certain social settings, such as small informal enterprises, are particularly likely to receive and pass on inaccurate information and are also more likely to engage in risky sexual conduct (Wang and Gao 2000). Peer education work, which is popular in China, has revealed that peer-based education programmes must take into account different sexual sub-cultures and use methods appropriate to each group (Wang and Van de Ven 2003). For instance, a storytelling tradition among trishaw-drivers has been used to successfully communicate information, thus bypassing the problems associated with illiteracy; activities such as the creation of drop-in centres are also important (Wang and Van de Ven 2003). Open discussions among young unmarried men also offer the potential to disseminate reproductive and sexual health information.

For Chinese teenagers the Internet has become a significant source of infor-mation (Wilson 2001), either replacing or filling in the gaps left by the lack of adequate education and parental communication and guidance (Brown and Stern 2002). However, the full extent and impact of the available information on a young audience must be further looked into, especially as such media can address taboo subjects in ways attractive to a young audience, but such a source of information can be also misleading or even dangerous (Brown and Stern 2002). Certainly many migrants in China's cities do use Internet cafes on their day off. Yet it is unknown just how widespread Internet use is among migrants. It is also unknown which sites they most commonly visit, and such information would be necessary in order to make sexual health content more effectively available to them.

Even though government programmes neglect to cater to the needs of the unmarried, market forces and opportunities for earning money mean that hospitals and clinics have started to run pharmacies which sell contraception. Since the early 1990s, shopping centres in even small urban areas have been selling all forms of contraception (Hyde 2000). Yet there are several factors that limit young migrant's access to these goods. One set of factors involves gendered assumptions and attitudes towards contraception. Young men have often been embarrassed to buy condoms and have found it convenient to assume that young women were taking the pill. Young women for their part have commonly relied on informal networks to get hold of the pill because long working hours and lack of money has often meant that they have been unable to get reliable access themselves. Furthermore, the decline of the socialist healthcare system and the rise of the private sector have put the cost of contraception beyond the means of many, who then end up not using any at all.

While the general tendency for Chinese scholarship has been to focus on sexual education and information for married migrants, some studies have looked at women's experiences more generally, or at the younger population, with revealing results. For example, a study of over 3,000 unmarried men and women in Beijing conducted in the 1990s showed that 47 per cent had had sexual relations before marriage (Jiang 1997). Findings from this and similar studies consistently show that migrants, in particular female migrants, lack reproductive and sexual health information or access to services offering choices about reproductive health and contraceptive use (Liu *et al.* 2004), even though a receptive audience for such facilities has been identified among both married and unmarried individuals (Z. Zhang 1999).

Maternity services for migrant women

Migrant women are particularly vulnerable when it comes to access to maternity services, and face difficulties in obtaining good quality maternity care regardless of whether they stay in the city or return to the village to give birth. Overwhelmingly, migrant women's inability to gain access to services, especially in urban areas, is linked to difficulties in obtaining access to medical insurance (Wang *et al.* 2007). Migrant women are also frequently denied maternity leave (Wong *et al.* 2007). The problems with services mean that women will often wait until late in their pregnancy to present for their first prenatal examination (F. Wang *et al.* 2005). Further, as already mentioned, migrant women are likely to have abortions while still young and to have abortions later during their pregnancies (Zheng and Lian 2005).

In cities, migrant women may lack access to personal postnatal care and rest because they are separated from traditional support networks (Chu 2005) and must resume busy work schedules (F. Wang *et al.* 2005). This may be experienced as a particular hardship as a result of customary expectations that women who have given birth are entitled to take a one-month resting period known as *zuo yuezi*. Women from rural areas may also wish to have traditional birth attendants, and doctors familiar with different notions about the transmission of diseases, which do not necessarily follow accepted western practices (JHUCCP 1996). The few migrants who are able to afford to give birth in cities may therefore encounter unfamiliar and even distressing environments.

The majority of migrant women who are forced to return to give birth in the countryside also experience difficulties (Wang 2006; Zheng and Lian 2005). This is because there is a serious shortage of appropriate medical care in rural areas. In discussions with women during fieldwork in northwestern China, the author learnt that no care was provided for women who gave birth at home in villages. In the event of complications during delivery, families would be faced with the difficulty of getting help in a situation where transport and communications were inadequate. Help, if and when available, is expensive. However, doctors in the local town hospital emphasised that they would travel out to a village to assist an expectant mother and remain there until the baby was born, even if delivery was delayed (Hoy 2004b). Despite their differences, both testimonies demonstrate the difficulties of providing acceptable pre-, neo- and post-natal care for a woman and her child in rural areas. Admittedly women may derive a certain benefit from returning home to have their baby and from the familiar and potentially supportive environment. The expectant mother may experience a psychological advantage that would be less available in the urban area. This does not, however, compensate for the failing health system in rural areas.

The role and work of non-governmental organisations

In response to migrants' urgent needs for sexual health services and education international and indigenous NGOs have carried out projects in partnership with each other and with the Chinese government. This section looks at the kind of programmes which have been running in China, and outlines the possible future direction of these programmes as well as lessons from HIV/AIDS programmes. Many of these programmes have been shaped by moral issues, but successful interventions reserve judgment and instead recognise the multiple threats faced by all migrants, especially women.

The China Youth Development Foundation and the Chinese Family Planning Association have run projects which have focused on adolescent sexual and reproductive health (Ru 2005). An innovative Chinese organisation, the Shanghai Family Planning Association (SFPA), has been experimenting with different methods in its work with adolescents. For instance, a Youth Healthy Living Project was organised in 2003 which offered 330 young unmarried individuals information and guidance on reproductive health issues. The SFPA was also involved in designing a website for Shanghai Jiaotong University that focused on AIDS prevention. Since the 1990s, an increasing number of Chinese NGOs, including the China Population Welfare Foundation and the Yunnan Reproductive Health Research Association, have increasingly worked with international NGOs, for instance Save the Children and the Australian Red Cross.

Marie Stopes China has been active in introducing sensitive interventions aimed at young people and the migrant population. An important feature of these projects is that they also provide support for teachers who have often not had any training in sexual and reproductive health education (Marie Stopes, n.d.). The Asia Foundation has run programmes specifically for migrant women workers in Guangdong. Here, education and training, preventive health services, legal advice and counselling have been offered free of charge to some 200,000 women (Asia Foundation 2004). The Ford Foundation, Swedish International Development Cooperation Agency (SIDA), UNICEF and United Nations Development Programme (UNPD)/UNAIDS are all increasingly active in providing sexual health education and services for migrants. The UNDP has suggested that multipartite HIV/AIDS prevention systems should be based at migrants' workplaces to ensure greatest effectiveness, and be embedded into existing anti-discriminatory work practices. Citing successful examples from the Philippines, UNDP has also strongly recommended that NGOs recognise, and work with, community-based migrant organisations so that migrant interests are better represented (UNDP 2003).

As noted above, the sexual and reproductive health education programmes often had at least an implicit moral dimension. More recent initiatives however have been more pragmatic and have explicitly avoided raising issues of moral blame (Hsu and du Guerny 2000). Yet this approach has not always been taken on board by the Chinese government which has tended to avoid a committed and constructive out-reach in such areas as HIV/AIDS in hard-to-reach populations like migrants (Thompson 2005), and has recently restricted the work of AIDS activists in China (Human Rights Watch 2005).

When successful, the support work carried out by NGOs recognises the vulnerable situation of women and their needs. Approximately half the women in the migrant population are under 25 years of age, single, and concentrated in small enterprises and the service industry (World Health Organisation 2002). The Department of Reproductive Health and Research at the World Health Organisation concluded that pro-active measures for sexual and reproductive health should reflect the fact that pre-marital sex is no longer rare, that it results in unwanted pregnancies and abortions, and that the lack of information about contraception is a serious concern. The department identified four obstacles to the use of contraception: unplanned sex, the wish to please partners, embarrassment, and the belief that family planning centres would not help unmarried women. All these issues are very relevant to the Chinese situation, and the Chinese government could look into the experience and information available worldwide for practical examples regarding the provision of appropriate support and education to particular population segments (World Health Organisation 2002).

An article in the China Development Brief reports on a new approach to delivering sex education in China which breaks through taboos (Qian 2005). This new programme, which has aimed to place sex within the wider context of normal human relationships, was introduced in a five-year project (2000–2005) promoted by the China Family Planning Association (FPA) and supported by an American non-profit organisation PATH (Programme for Appropriate Technology in Health). The project has been launched on a trial basis in cities including Beijing, Tianjin and Shanghai with the cooperation of municipal education and propaganda authorities. PATH provided comprehensive guidance in life skills to migrants working for a large company with a network of branch offices across China, resulting in a 35 per cent drop in unplanned pregnancies (Programme for Appropriate Technologies in Health n.d.).

Flexible and sensitive sexual and reproductive health education programmes for young people have been launched on a trial basis in China since the mid-1990s and early 2000. The author attended an early peer education project in one of Shanghai's medical universities in which first-year medical students received basic biological information, were introduced to the contraceptive pill and shown how to use a condom (Hoy 2001b). While many organisations and programmes have been successful in their own area of responsibility and activity, information is not being shared among them and the extent to which they are successful in challenging established taboos in the society at large, and not only within the groups included in such projects, is not known. This issue is discussed in greater depth in the next section.

Other examples of innovative approaches to the provision of sexual and reproductive health education and support include a new strategy for reproductive health developed by the Family Planning Association in 2004, along the lines of a UNFPA-funded reproductive health programme for rural areas (Qian 2005). In the same year, Marie Stopes China implemented a project aimed at junior high-school children. Other organisations are working to delay the age of sexual initiation or resolve minor health problems (China Development Brief 2000). For many the prevention of HIV/AIDS infections is an important priority (Save the Children 2004–05).

As HIV/AIDS programmes in China continue to suffer from competition for funds, project replication, lack of communication and the failure to develop networking capacity, a Hong Kong-based charity has developed an HIV/AIDS programme directory intended to forge greater coherence and cooperation among the many different programmes. The first China AIDS Info directory was published in 2005 and annual updates are planned (China AIDS Info 2005). It lists nearly 180 NGOs, UN agencies, bilateral programmes, Chinese government agencies and private sector groups offering HIV/AIDS services. A similarly chaotic pattern of programmes is repeated in relation to reproductive and sexual health education issues. It remains to be seen whether AIDS Info can stimulate better cooperation within this community.

Remaining challenges

The ongoing reform programme in China has generated massive economic growth, but has also resulted in social inequity and the marginalisation of certain groups. Yet there has been no support for the familial relationships on which such vulnerable groups have to rely. Social welfare reform has so far concentrated only on the issue of social security (Shang *et al.* 2005). To be effective, public health interventions, which are central to family as well as individual capacity and wellbeing, must be part of such a wider approach to reform that targets all vulnerable groups.

The decline in the overall status of women in China, especially of rural women who are at the bottom of the social scale, is of particular relevance to this discussion. There is a need for the government to acknowledge that migrants are not simply a transient phenomenon among urban communities, but that, given the opportunity, they can make a positive contribution through their labour and be committed members of urban society. As such, their presence and contribution should be reflected in urban and social planning programmes and included at all levels of governance (Wu 2004).

Though aid from international donors has been forthcoming, there is general awareness and recognition of the problems created by the near-exclusive reliance on projects sponsored by donors, NGOs and the government (Jing 2004). The utility and lasting impact of even well-designed projects and appropriate services will not be assured unless they are provided on a sustainable and comprehensive basis and reach all sectors of society (Ru 2005; Thompson 2004; Zheng and Lian 2005). Current programmes fall short of this as they fail to upgrade practices or to promote greater capacity building (Ru 2005; Thompson 2004).

The benefits of involving a wide spectrum of civil society representatives, such as NGOs, the media, the business and corporate sector, academics and women themselves in making services available to the population in general, including migrants, and pressuring the government to develop and improve the relevant policies, services and rights, have gained renewed interest (Thompson 2004; Zhang 2003). However, the growing presence and involvement of civil society is also a cause for some concern on the part of the Chinese government because it provides scope for activities it does not necessarily sanction. Yet while the importance of civil society in working with communities to help develop best practices is undoubtedly valuable (Kaufmann and Jing 2002; Zhang 2003), how this is to occur in the Chinese context is less clear.

There are contrasting views on budgetary control over programmes and policies, as well as their expected impact. Some argue that decentralised economic support would enhance the capacity of local grassroots organisations to target their activities more effectively (Thompson 2004). In contrast, others feel that financial devolution would not contribute to the creation of good programmes and argue for extra-budgetary resources (Kaufmann and Jing 2002). Moreover, it is important that services at the local level intended for particularly vulnerable groups, such as migrants, are provided free of charge (Liu *et al.* 2004).

Many remedial measures designed to produce an integrated, comprehensive system of reproductive health and sexual education and support are avai-lable and include culturally appropriate education and prevention programmes (Laumann and Mahay 2002; Liu *et al.* 2004; Wang and Gao 2000; Wang and

Van de Ven 2003). However, for programmes to be effective and as far as possible self-sustaining, such measures must address such issues as the curriculum, programme methods, support for educators and a realistic timeframe. Integrated programmes should ideally include the whole community and take into account differences within communities, for instance the presence of particularly high-risk groups. Established networks, such as family planning services, could be more effectively included in projects.

The staff must understand the nature of long-term impacts aimed for, and acknowledge them in the programmes to be conducted. This is especially important where education programmes aim to influence and change behaviour, and not just to inform. For example, the age of sexual initiation is important as it will affect later adult behaviour and outcomes (Laumann and Mahay 2002). Thus, programmes looking at early sexual behaviour need to take into account long-term behavioural influences on later adult behaviour. Depending on the type of programmes, they will be confronted with different challenges. Projects which aim to enhance primary health, sexual and reproductive education are significantly more difficult to implement than those aiming to limit the fertility rate (Harcourt 1997b).

As we have seen, new policies by themselves, however comprehensive, are not sufficient to bring about a change in social attitudes and behaviours (Kirby *et al.* 2005). Policy reform and national leadership are needed at all levels, including the local level. Yet, weak or absent overall leadership strategies in China (Ru 2005) are a cause for concern. Appropriate national HIV/AIDS programmes have yet to be established and those that exist frequently exclude young people and migrants and are hampered by the lack of effective monitoring and evaluation programmes and by the failure to share information (Ru 2005). There is some concern that the establishment of equally important national programmes for other sensitive issues or populations, such as migrants and their needs, might also be neglected (Thompson 2004). As already mentioned, the Asia Foundation is trying to initiate activities that will remedy at least the problem of information sharing for those involved in supporting the migrant community, and intends to create an information exchange network to collect and publicise project information and results and, significantly, to disseminate good practice and prevent project duplication (Asia Foundation 2004).

The implementation of programmes to provide reproductive and sexual health services and education for migrants has direct implications for China's achievement of key Millennium Goals (Usher 2005). Only by promptly acknowledging that the migrant population is a valuable and fundable part of the wider national community and that they have wide-ranging and multi-faceted human needs can comprehensive policies be applied which address the promotion of maternal health (Millennium Goal 5), gender equality and the empowerment of women (Millennium Goal 3). Both these MDGs imply benefits for society at large, and only by taking into account the legitimate and complex needs and rights of migrant individuals, can their vulnerability and marginalisation be addressed.

References

Ahmadi, N. (2003) 'Migration challenges views on sexuality', *Ethnic and Racial Studies*, 26(4): 684–706.

Asia Foundation (2004) 'Programmes to support migrant women workers', Report for the Asia Foundation, Beijing. Available at: http://www.asiafoundation.org

Bannister, J. (1987) *China's Changing Population*. Stanford: Stanford University Press.

Berer, M. (ed.) (1998) 'Women's health services: where are they going?' *Reproductive Health Matters*, 6(11): 10–12.

Brown, J.D. and Stern, S.R. (2002) 'Mass media and adolescent female sexuality', in Wingood, G.M. and DiClemente, R.J. (eds) *Handbook of Women's Sexual and Reproductive Health*. New York: Kluwer Academic and Plenum Publishers, pp. 93–112.

Buitelaar, M.W. (2002) 'Negotiating the rules of chaste behaviour: reinterpretations of the symbolic complex of virginity by young women of Moroccan descent in the Netherlands', *Ethnic and Racial Studies*, 25(3): 462–489.

Buston, K. and Wight, D. (2006) 'The salience and utility of school sex education to young men', *Sex Education*, 6(2): 135–150.

China Development Brief (2000) 'Reproductive health education project builds on peer methods', January. Available at: http://www.chinadevelopmentbrief.com/node/233, accessed 1 April 2006.

Chen, Y. and Zheng, G. (2001) 'Butong shenghuo gongzuo tiaojianxia nuxing liudong renkou shenzhi jiankang qingquang de diaocha he sikao' (Investigation and reflection on the status of the female floating population's reproductive health in different living and working environments), *Nanfang renkou* (South China Population), 3(16): 53–59.

China AIDS Info (2005). Available at: http://www.china-aids.org

Chu, C.M.Y. (2005) 'Post-natal experience and health needs of Chinese migrant women in Brisbane, Australia', *Ethnicity and Health*, 10(1): 33–56.

Easterly, W. (2007) 'How, and how not, to stop AIDS in Africa: Essay on "The invisible cure: Africa, the West and the fight against AIDS" by Helen Epstein', *The New York Review of Books*, 16 August, LIV, (13): 24–26.

Evans, H. (1997) *Women and Sexuality in China*. Cambridge: Polity Press.

Farrer, J. and Sun, Z. (2003) 'Extramarital love in Shanghai', *The China Journal*, 50(July): 1–36.

Farquhar, J. (2002) *Appetites: Food and Sex in Post-Mao China*. Durham and London: Duke University Press.

Friedman, S.L. (2000) 'Spoken pleasures and dangerous desires: sexuality, marriage and the state in rural south-eastern China', *East Asia*, 18(4): 13–27.

Gardner, R. and Blackburn R. (1996) 'People who move: new reproductive health focus', *Population Reports, Series J, No. 45*, Johns Hopkins School of Public Health, Population Information Programme, Baltimore, November.

Greenhalgh, S. and Winckler, E.A. (2005) *Governing China's Population: from Leninist to Neoliberal Biopolitics*. Stanford: Stanford University Press.

Harcourt, W. (1997a) 'Conclusions: moving from the private to the public political domain', in Harcourt, W. (ed.) *Power Reproduction and Gender: the Intergenerational Transfer of Knowledge*. London and New Jersey: Zed Books, pp. 184–197.

Harcourt, W. (1997b) 'An analysis of reproductive health: myths, resistance and new knowledge', in Harcourt, W. (ed.) *Power Reproduction and Gender: the Intergenerational Transfer of Knowledge*. London and New Jersey: Zed Books, pp. 8–34.

Hewamanne, S. and Müller, H.P. (1997) 'The impact of the global economy: returnee migrant workers in Sri Lanka' in Harcourt, W. (ed.) *Power Reproduction and Gender: the Intergenerational Transfer of Knowledge.* London and New Jersey: Zed Books, pp. 120–138.

Hoy, C.S. (2001a) 'Adolescents in China', *Health and Place* 7(4): 261–271.

Hoy, C.S. (2001b) Fieldwork, Shanghai, April.

Hoy, C.S. (2004a) Fieldwork, Shanghai, October–November.

Hoy, C.S. (2004b) Fieldwork, Xinjiang, February.

Hoy, C.S. (2006) 'Learning about sexual and reproductive health issues: young migrant women's experiences in China', paper presented at the International Workshop on Sexuality and Migration in Asia, National University of Singapore, Asian MetaCentre for Population and Sustainable Development Analysis, Asian Research Institute, April 10–11.

Hoy, C.S. (2007) 'Migration as sexual liberation? Examining the experience of young female migrants in China', *Children's Geographies* 5(1–2): 183–187.

Human Rights Watch (2005) 'Restrictions on AIDS Activists in China'. Available at: http://hrw.org/reports/2005/China0605/3.htm, accessed 25 January 06.

Hsu, L.N. and du Guerny, J. (2000) 'Population movement, development and HIV/AIDS: looking towards the future', report for the United Nations Development Programme, Thailand.

Hyde, S.T. (2000) 'Selling sex and sidestepping the state: prostitutes, condoms and HIV/AIDS prevention in southwest China', *East Asia* 18(4): 108–124.

Huang, P. and Zhan, S. (2005) 'Internal migration in China: linking it to development', paper presented at the Regional Conference on Migration and Development in Asia, Lanzhou, China, 14–16 March. Available at: http://www.iom.int/chinaconference/files/documents/bg_papers/China.pdf

Jefferies, E. (2004) *China, Sex and Prostitution.* London and New York: Routledge Curzon.

Jiang, S. (1997) 'Jiushi niandai chengshi weihun qingnianxing guannian, xingxing diaocha' (A survey of attitudes and behaviour towards sexuality of urban unmarried youth in the 1990s), *Zhongguo Renkou Kexue* (Population Science of China), 2: 45–53.

Jing, F. (2004) 'Health sector reform and reproductive health services in poor rural China', *Health Policy and Planning*, 19 (Suppl. 1): 140–149.

Johns Hopkins University Center for Communication Programs (JHUCCP) (1996) 'People who move: new reproductive health focus'. Available at: http://www.jhuccp.org/pr/j45

Kaufmann, J. and Jing, J. (2002) 'China and AIDS – the time to act is now', *Science*, 296(June): 2339–2340.

Kirby, D, Laris, B. and Rolleri, L. (2005) 'Impact of sex and HIV education programmes on sexual behaviours of youth in developing and developed countries', youth research working paper no. 2, Family Health International, Research Triangle Park.

Laumann, E.O. and Mahay, J. (2002) 'The social organisation of women's sexuality', in *Handbook of Women's Sexual and Reproductive Health.* New York: Kluwer Academic/Plenum Publishers, pp. 43–70.

Liang, Z. and Ma, L. (2004) 'China's floating population: New evidence from the 2000 census', *Population and Development Review*, 30(3): 467–488.

Liu G. (1997) 'An investigation of adolescent health from China', *Journal of Adolescent Health*, 20(4): 306–308.

Liu, H. *et al.* (2004) 'Liudong renkou de shengzhi jiankang fuwu' (Migrants' reproductive health services), *Renkou Yanjiu* (Population Research), 25(5): 92–96.

Liu, H., Li, X., Stanton, B., Liu, H., Liang, G.J., Chen, X.G., Yang, H.M. and Hong, Y. (2005) 'Risk factors for sexually transmitted disease among rural-to-urban migrants in China: implications for HIV/sexually transmitted disease prevention', *AIDS Patient Care and STDs*, 19(1): 49–57.

Lou, B., Zheng, A., Connelly, R. and Roberts, K.D. (2004) 'The migration experiences of young women from four counties and Sichuan and Anhui', in Gaetano, A. and Jacka, T (eds) *On the Move: Women in Rural-to-Urban Migration in Contemporary China*. New York: Columbia University Press, pp. 207–242.

Macmillan, J. (2004) 'Doing it by the book: natural tales of marriage and sex in contemporary Chinese marriage manuals', *Sex Education*, 4(3): 203–215.

Marie Stopes (n.d.) Marie Stopes China. Available at: http://www.youandme.net.cn

Murphy, R. (2004) 'The impact of labour migration on the well-being and agency of rural Chinese women: cultural and economic contexts and the life course', in Gaetano, A. and Jacka, T (eds) *On the Move: Women in Rural-to-Urban Migration in Contemporary China*. New York: Columbia University Press, pp. 243–276.

Parish, W.L. *et al.* (2003) 'Population-based study of Chlamydial infection in China: a hidden epidemic', *Journal of the American Medical Association*, 289(10): 1265–1273.

Parish, W.L. (2002) 'Open door sexuality', *University of Chicago Magazine*. Available at: http://magazine.uchicago.edu/0210research/invest-open.html, accessed 24 January 2006.

People's Republic of China Law on Population and Birth Planning (2002), Chapter 6, Supplementary Provision, Article 45, passed by the 25th Meeting of the Standing Committee of the Ninth National People's Congress on 29 December 2001. Available at: http://www.unescap.org/esid/psis/population/database/poplaws/law_china/ch_record052.htm#chapter6

Pitanguy, J. and de Mello e Souza, C. (1997) 'Codes of honour: reproductive life histories of domestic workers in Rio de Janeiro', in Harcourt, W. (ed.) *Power Reproduction and Gender: the intergenerational transfer of knowledge*. London and New Jersey: Zed Books, pp. 72–97.

Popinfo (2003) 'Liudong renkou jihua shengy guanli he fuwu gongzuo guiding', (Regulation on the administration and work of family planning in the floating population), Guojia renkou he jihua shengyu weiyuanhuiling (State Population and family planning committee), 15 December 2003. Available at: http://www.popinfo.gov.cn/popinfo/pop_doczcwd.nsf/, accessed 13 February 2006.

Programme for Appropriate Technologies in Health (PATH) (n.d.) 'China Adolescent Health Project Report: Meeting young people on the move'. Available at: http://www.path.org/projects/china_adolscent_healt_project.php, accessed 8 February 2006.

Puri, M.C. and Busza, J. (2004) 'In forests and factories: sexual behaviour among young migrant workers in Nepal', *Culture, Health and Sexuality*, 6(2): 145–158.

Qian, T. (2005) 'Sex education begins to break taboos', *China Development Brief*, May 8. Available at: http://www.chinadevelopmentbrief.com/node/57, accessed 4 January, 2006.

Ru, X. (2005) Youth and HIV/AIDS in China. Available at: http://www.ksg.harvard.edu/cbg/asia/HIVAIDS%20papers16%-Youth%20and%20HIV.pdf., accessed 19 February 2006.

Save the Children (2004–2005) 'Country brief: China'. Available at: www.savethechildren.org.uk.

Shanghai Family Planning Association (2004) 'Shanghaishi "qingchun jiankang" xiemutongxu' (The messages of adolescent reproductive health programmes in Shanghai) VII(1), Shanghaishi jihuashengyu banxie qingchunqiankang xiemu bangongshi,

Shanghai Family Planning Association Adolescent Reproductive Health Programme Office, Shanxinanlu, Shanghai.

Shang, X., Wu, X. and Wu, Y. (2005) 'Welfare provision for vulnerable children: the missing role of the state', *The China Quarterly* 181: 122–136.

Smith, C.J. (2005) 'Social geography of sexually transmitted diseases in China: exploring the role of migration and urbanisation', *Asia Pacific ViewPoint*, 46(1): 65–80.

State Council (1998) 'Measures on the administration of family planning for the floating population', approved by the State Council on August 6, 1998, effective 1 January 1999. Available at: http://www.unescap.org/esid/psis/population/database/poplaws/law-china/ch_record0, accessed 13 February 2006..

State Council (2003) 'Guo wu yuan ban gong ting guan yu zuo hao nong min gong jin cheng wu gong jiu ye guan li he fu wu gong zuo de tong zhi' (The State Council's document on offering better management and services for peasants who migrated to work in cities). Available at: http://www.gov.cn/zwgk/2005-08/12/content_21839.htm

Thompson, D. (2004) 'China faces challenges in effort to contain HIV/AIDS crisis', *The Population Reference Bureau*. Available at: http://www.prb.org/Template.cfm?Section=PRB&template=/ContentManagement/Co, accessed 7 September 2005.

Swedish Agency for International Development Cooperation (SIDA). Available at: http://www.sida.se

Tsukamoto, K. (2005) 'Family ties: there's a hidden cost to China's migration', The Asahi Shimbun. Available at: www.asahi.com/english/world/TKY20050305151.html, accessed 19 April 2006.

UNICEF (2007) www.unicef.org/china/children_877.html, accessed 21 May 2007.

United Nations Development Programme (UNDP) (2003) Second Meeting Report, United Nations Regional Task Force on Mobility and HIV Vulnerability Reduction Zhengcheng, People's Republic of China.

United National Economic and Social Council (2002) 'Concise report on world population monitoring: reproductive rights and reproductive health with special reference to HIV/AIDS'. UNESC, E/CN.9/2002/2, pp. 6–12. Available at: http://www.un.org/documents/ecosoc/cn9/2002/english/ecn92002-2.pdf, accessed 20 September 2005.

Usher, E. (2005) 'The Millennium Development Goals and Migration', IOM Migration Research Series, International Organization for Migration, Geneva.

Wang, B., Hertog, S., Meier, A., Lou, C. and Gao, E. (2005) 'The potential of comprehensive sex education in China: Findings from suburban Shanghai', *International Family Planning Perspectives*, 31(2): 63–72.

Wang, B., Li, X., Stanton, B., Fang, X., Liang, G., Liang, G., Liu, H. and Yang, H. (2007) 'Gender differences in HIV-related perceptions, sexual risk behaviours and history of sexually transmitted disease clinics', *AIDS Patient Care and STDs*, 21(1): 57–68.

Wang, F., Ping, R., Zhan, S., and Shen, A. (2005) 'Reproductive health status, knowledge and access to health care among female migrants in Shanghai, China', *Journal of Biosocial Science*, 37(5): 603–622.

Wang, J. (2006) 'Health sector reform in China: gender equality and social justice', in Razavi, S. and Hassim, S. (eds) *Gender and Social Policy in a Global Context: Uncovering the Gendered Structure of 'the Social'*. Palgrave: Basingstoke, pp. 258–277.

Wang, S. and Van de Ven, P. (2003) 'Peer HIV/AIDS education with volunteer trishaw drivers in Yaan, People's Republic of China: process evaluation', *AIDS Education and Prevention*, 15(4): 334–345.

Wang, S.M. and Gao, M.Y. (2000) 'Employment and contextual impact of safe and unsafe sexual practices for STI and HIV: the situation in China', *International Journal of STD and AIDS*, 11(8): 536–544.

White Paper (2005) 'Government White Paper VI: Women and Health'. Available at: http://www.china.org.cn/e-white/20050824/6.htm, accessed 28 November 2005.

Wilson, S.N. (2001) 'Learning about sexuality from the internet', *Independent School*, 60(2): 50–55.

Wolfers, I., Fernandez, I., Verghis, S. and Vink, M. (2002) 'Sexual behaviour and vulnerability of migrant workers for HIV infection', *Culture, Health and Sexuality*, 4(4): 459–473.

World Health Organization (2002) 'Young female migrant workers in China in need of reproductive health information and services', *Social Science Research Policy Briefs*, 2(2), May, UNDP/UNFPA/WHO/World Bank Special Programme of Research, Development and Research Training in Human Reproduction, Department of Reproductive Health and Research, World Health Organization, Geneva.

Wu, F. (2004) 'Urban poverty and marginalisation under market transition: the case of Chinese cities', *International Journal of Urban and Regional Research*, 28(2): 401–423.

Xiang, B. (2004) 'Migration and health in China: Problems, obstacles and solutions', Asia MetaCentre Research Paper Series 17, Asia Research Institute, National University of Singapore.

Xu, F. (2000) *Women Migrants in China's Economic Reform*. Basingstoke: Macmillan.

Yang, H., Li, X., Stanton, B., Chen, X., Liu, H., Fang, X., Lin, D. and Mao, R. (2005a) 'HIV-related risk factors associated with commercial sex among female migrants in China', *Health Care for Women International*, 26(2): 134–148.

Yang, H., Li, X. Stanton, B., Fang, X., Lin, D., Mao, R., Liu, H., Chen, X. and Severson, R. (2005b) 'Workplace and HIV related sexual behaviours and perceptions among female migrant workers', *AIDS Care*, 17(7): 819–833.

Yang, X. (2000) 'The fertility impact of temporary migration in China: a detachment hypothesis', *European Journal of Population*, 16(2): 163–183.

Yang, X. (2004) 'Temporary migration and the spread of STDs/HIV in China: is there a link?', *International Migration Review*, 38(1): 212–235.

Yang, X. (2005) 'Does where we live matter? Community characteristics and HIV and sexually transmitted disease prevalence in south-western China', *International Journal of STD and AIDS*, 16(1): 31–37.

Zeng, B., Gui and Yu (2005) 'Liudong renkou shengzhi jiankang qingkuang fenxi' (Analysis of migrants' reproductive health situation), *Xiandai Yufang Yixue* (Modern Preventive Medicine), 32(1): 52–53.

Zha, B. and Geng, W. (1992) 'Sexuality in urban China', *Australian Journal of Chinese Affairs*, 28: 1–20.

Zhang, F. *et al.* (2002) 'Wuxi shi liudong renkou shengzhi jiankang qingkuang diaocha' (Wuxi city survey of the floating population's reproductive health situation), *Zhongguo fuyou baojian* (China maternal and child hygiene), 17: 228–229.

Zhang, H.X. and Locke, C. (2002) 'Contextualising reproductive rights challenges, the Vietnam situation', *Women's Studies International Forum*, 25(4): 443–455.

Zhang, K. and Li, D. (1999) 'Changing attitudes and behaviour in China: implications for the spread of HIV and other sexually transmitted diseases', *AIDS Care*, 11(5): 581–589.

Zhang, W. (2003) 'Pay special attention to adolescent sex and reproductive health and rights', speech by the Minister of the State Population and Family Planning

Commission to mark World Population Day, July 11. Available at: http://www.cpirc.org. cn/en/enews20030721.htm, accessed 7 September 2005.

Zhang, Y. (2002a) 'Migrant women workers and the emerging civil society in China', *The Asia Foundation Publications*. Available at: http://www.theasiafoundation.org/ publications

Zhang Y. (2002b) 'Hope for China's migrant women workers', *China Business Review*. Available at: http://www.chinabusinessreview.com/public/0205/ye.html

Zhang, Z. (1999) 'Liudong renkou zhong yuling funu shengzhi jiankang de zhuangqing yu yikao' (Reflections on the current situation of reproductive health of women of reproductive age in the floating population), *Renkou XueKan* (Population Journal), 4: 55–59.

Zheng, T. (2004) 'From peasant women to bar hostesses: gender and modernity in post-Mao Dalian', in Gaetano, A. and Jacka, T. (eds) *On the Move: Women in Rural-to-Urban Migration in Contemporary China*. New York: Columbia University Press, pp. 80–108.

Zheng, X. (1997) 'A survey of graduate students' knowledge, views and behaviour with respect to reproductive health', *Chinese Journal of Population Science*, 9(2): 123–133.

Zheng, Z. and Lian, P. (2005) 'Health vulnerability among temporary migrants in urban China', paper presented at the XXV International Population Conference, 18–23 July, Institute of Population and Labour Economics, Chinese Academy of Social Sciences, China. Available at: http://IUSSP2005.princeton.edudownload.aspx? submissionId=51980070905.

6 Housing and migrants in cities

Ya Ping Wang and Yanglin Wang

Introduction

Providing housing for migrants who live in cities is a challenge that faces all developing countries. In China more than 100 million migrants live in cities and towns. So far, the government has made no major effort to provide affordable housing for them. At the same time, China's cities do not have the kinds of large-scale slum settlements that are found in the cities of other developing countries. This, however, does not mean that China has solved the housing problems of migrants. To the contrary, most migrant workers live in poor conditions either at their work sites or in low-quality housing concentrated at the urban–rural interface zones. It is very common for several migrants or families to share a room. Essential facilities such as water, electricity and toilets in migrant areas are usually poor quality. Migrant labourers are classified as temporary or 'floating' and receive negligible social or economic support from municipal governments even though they provide the labour necessary for rapid urban economic growth. Most poignantly, they have no access to the new houses that they help to build.

Other chapters in this volume and studies published elsewhere have discussed such issues as migrants' demographic characteristics, the magnitude of mobility, the composition of migration streams and regional patterns of migration (see Ma and Xiang 1998; Davin 1999; Knight and Song 1999; Solinger 1999; Fan 2001; Goodkind and West 2002; Hussain 2002; Murphy 2002; Wu 2002). This chapter describes the housing conditions of migrant workers. It draws on recently published works in both Chinese and English and on our own research on this subject conducted over the past several years. Our chapter focuses on the location and characteristics of migrant housing as well as on rent and other living costs. We also discuss recent policy developments and management arrangements related to migrant housing.

The tradition of migration

Migration has been one of the most important human development activities in China. During the first half of the twentieth century, rural people viewed migration to towns and cities as a household and personal development strategy. There was a saying among the members of the older generations in villages: *nan zhang shi er' tuo fu zi* (boys leave their home and parents at twelve to develop a career). Many young men of our parents' generation went to towns to work in shops, restaurants and other commercial establishments. They started as junior assistants (apprentices) helping bosses to run their business. Work could include making tea, fetching water and making beds. They also learnt to read and write through accounting and book-keeping. This was their education; there was no school in the village and not many families could afford to send their children to the town for education. Boys who were brave enough to go out and find a job in the town were *you chu xi* (clever and with a bright future), while those who stayed at home were *mei you chu xi* (without promise).The character *chu* means 'going out'.This tradition of migration continued after 1949 when the Communists established the People's Republic of China. Those who went to the towns and cities found jobs in the new government and became government officials and subsequently, after the *hukou* system was introduced, they became urban residents. Their courage, open mindedness, and reading and writing skills (literacy) gave them an advantage. Those who stayed at home became farmers and were later registered as members of the rural population.

Since 1949, government policy has been an important determinant of migration patterns. During the first few years of the Communist government there were no formal controls on migration, and millions of rural labourers rushed into new industrial cities or towns for employment. Xian city, for example, saw a dramatic increase in its population from around 300,000 in 1949 to over one million by 1957 (Wang and Hague 1992). This large-scale increase in the size of the urban population prompted the introduction of the *hukou* registration system, under which any move from a rural to an urban area had to be approved by the government.

Recent migration studies pay particular attention to the large-scale movement of population from rural areas to towns and cities during the period of the economic reforms which started in 1978. On account of scholars' focus on the reform era, however, the long tradition of migration and the adventurous nature of the Chinese people, which was in evidence prior to the reforms, have not received sufficient emphasis. Even under the *hukou* system, only a small portion of families lived peacefully at their registered homes. During the peak of the Cultural Revolution period, migration was common in both urban and rural areas. The movement of the Red Guards throughout the country and the subsequent rustication of urban youth, professionals and officials to rural areas was a reversal of previous rural–urban migration. Moreover, during the early 1970s, within rural areas, some large-scale temporary migration was organised by the government. Many rural labourers left their homes to work on major irrigation (*xiushuili*) and infrastructure projects. They either lived in the spare rooms of the local people or in temples, warehouses or local school classrooms. Usually there were no beds. Men and women occupied separate rooms. They slept in rows on the floor with only some straw under their quilts. They lived as if in a temporary army camp and their meals were cooked in temporary kitchens. Yet even though the work was hard and physically demanding, they received only a few *yuan* per month in pocket money. Most of their payment took the form of workpoints which were allocated in a manner similar to that used to remunerate farmers who worked back at home on the production team's land.

Subsequent migration during the reform era was of course substantially greater in volume to that which had occurred in the pre-reform era, but it was similarly characterised by migrants having to 'make do' with temporary living arrangements. Even after the 1984 regulations, which for the first time permitted rural people to work and live in towns, it was clear that these newcomers had to meet their own needs in that they were required to bring their own food supply with them (Xu 2001). In the following year, rural migrants were allowed to register as temporary residents in cities (Shen 1995). Rural–urban migration has since become a central part of urban development (Knight and Song 1999). During the 1990s, large-scale property and infrastructure development projects in the cities attracted many young people from the countryside to work. New industrial factory zones and market places for cloth, fresh food and vegetables were other city arenas which also drew in large numbers of rural migrants. Regardless of where the migrants work, however, poor living conditions have been a common feature of their urban-based lives.

Location of migrant housing

When contemplating the housing requirements and preferences of migrants, it is helpful to keep in mind differences and shifts in the identities and backgrounds of the people who have migrated. In the 1980s and early 1990s, migrants in Chinese cities referred mainly to those labourers who originated from rural areas. Urban officials who moved between cities with government approval were not normally counted as migrants. Over the past few years, however, owing to an increase in urban–urban migration, the composition of the migrant population has become more complex. In many coastal cities, the numbers of international and urban–urban migrants have been increasing. In particular, intensifying competition in the urban job markets has caused many college and university graduates to leave their home cities and search for better-paid employment in coastal areas.

The different backgrounds of the migrants affect their housing requirements. Highly paid international migrants live in rich areas or gated communities. University graduates who have moved from other cities (mainly inland cities where job opportunities were rare) tend to rent temporary accommodation while they look for a job. Then, once they have found a job and earn a more stable income, they move into mainstream housing areas, often purchasing their own place and integrating with the locally registered urban *hukou* residents. Rural and poor urban–urban migrants tend to concentrate in special areas. A distinction could be made between those living on construction sites and those living in housing that is not related to their work. Construction workers and industrial workers of large factories tend to live collectively in dormitories, usually provided by their employers. Other migrants have to find their own accommodation in the private rental sector. The most common locations for migrant housing include traditional old housing areas in the inner cities, rundown old housing estates associated with earlier industrial factories, and the so-called 'urban villages'.

Construction sites are one of the main locations where rural migrants live. Owing to large-scale property and infrastructure developments in cities, many migrants are employed in construction, refurbishment and decoration. One report estimates that in 2002 in Beijing, approximately 20 per cent of migrants were living on construction sites (Beijing Municipal Bureau of Statistics 2003). Another report estimates that in 2004, in Shenzhen, over 70 per cent of construction workers were poorly educated rural migrants. The average wages of these construction workers range from 1,000 to 1,500 *yuan* per month, well below the city average of 2,263 *yuan* per month (Shenzhen Municipal Labour and Social Security Bureau 2005). Many construction workers live in temporary shelters in the vicinity of the worksites, which are usually provided free of charge by the construction companies. Overcrowding is normal with average personal floor space ranging from 3–4 square metres, not much larger than the space of a single bed. A room of 10–30 square metres is typically shared by 6–20 people. Most construction workers are married men who leave their wife and children in the villages. They commonly share a room with several other men. Because these men are very busy and work long hours, they often lack private cooking equipment. Instead they eat in canteens run by the company. They also have access to a shared shower room and toilet, and some employers provide a TV and activity room.

The living arrangements of refurbishment and decoration workers are similar to those of construction workers. The size of the refurbishment teams varies substantially. On large projects, the teams often form part of a major construction workforce; for smaller projects such as family house decorating, they tend to work in a small group. The refurbishers work most of the time inside buildings where they are invisible to the public. The life of one group of workers who we met in Beijing provides insight into their living conditions. The group consisted of approximately 25 men aged between 20 and 40 years from Shandong province. They worked for a leader who scouted for jobs in the city. At the time of interview, their main task was to refurbish an old building. The workers brought their own quilts and slept on wooden boards inside the building they were refurbishing. Although the living conditions were basic and the smell of paint very strong, the accommodation was convenient and free of charge. Food was arranged by the team leader and brought to the site in a big container. They worked about 14 hours each day from 5 am to 10 pm with occasional short breaks. There were no weekends or holidays. Since the leader was from the same rural area, there was no contract between these workers and their leader, which also gave them flexibility in case they needed to return home for any reason. After food costs, they could earn about 700 *yuan* each month (the average salary in the city in 1999 was 1,148 *yuan* per month) which was considered a decent amount. They travelled from job to job in the city carrying only their quilt and a few personal belongings.

In the industrial areas of fast-growing coastal cities, some factories provide dormitories for their production line workers. In Dongguan city, for example, factory owners have built specifically designed multi-storey dormitories for the workers which are located next to the factory workshops. The living conditions of the factory workers are in general slightly better than those of construction workers. While almost all construction workers are male, factories employ a very high proportion of young females. In 2004 the Shenzhen municipal government carried out a survey of 25 large factories located in the city. The survey found that the average number of workers in these factories was 3,750. Over 60 percent of them were female, and 85 percent were younger than 30 years. The educational background of factory workers was slightly higher than that of construction workers. Their monthly income was, however, well below the city average, ranging from 600 to 1,500 *yuan*. Of the 25 factories surveyed, 21 provided dormitories for their production line workers. More than half of these dormitories were situated inside the factories (Shenzhen Municipal Housing System Reform Office 2004). Dormitory allocation normally reflects the grades of employees. For managerial-level employees, apartments were sometimes provided. For ordinary production line workers, dormitory rooms ranging from 20 to 40 square metres were shared by 6 to 10 people. Facilities in dormitories were very basic and bunk beds were common. Apart from beds, there was usually a washing area and toilet in each room. Factory workers seldom cooked for themselves and instead bought food from the factory-run canteen. A few factories even provided shopping, sport and recreation facilities including reading and TV rooms.

Other migrants, not employed by large organisations, live mainly in private rental housing in poor areas in cities. The location and conditions of these areas vary from city to city. They usually fall into one of two types: traditional and old housing areas in the city centre and farmers' housing in suburban villages. Property boom and urban redevelopment programmes have renewed much of the old town areas of most large cities. However, in almost every city, there are pockets of traditional housing areas left by the property developers. These pockets are usually small and difficult pieces of land situated on steep slopes, along railway lines and in between large organisations. Houses in these areas are mostly owned privately by local residents. The better-off original residents have moved to new houses in other areas, but they keep the old houses to secure compensation when the areas are redeveloped. These houses or rooms are rented out to migrants. A few years ago, particularly in large inland cities, these areas were prime locations for rural migrants. They normally had a long history of poverty before the communist period. The original density of houses in these areas was relatively low and there were sizable open spaces between rows of housing. Because of population increases, every family made additions to the original structure to maximise indoor space. The open spaces between buildings were gradually covered up. When we visited the city of Chongqing in 2001 we saw that the migrant workers mixed together with local poor families in extremely poor quality housing built on steep slopes. Though initially not of interest to commercial property developers, these areas were nevertheless very vibrant in economic terms, hosting small shops and street markets. The infrastructure in these localities was, however, very poor, with water pipes running along the street and sewage flowing out into open or covered up ditches. The majority of families did not have internal kitchens and toilets; and many families cooked their food on stoves outside the house. Most of the migrant residents either worked as *bangbang jun*[1] or as small traders selling farm products (Wang 2004).

The most prominent areas occupied by migrants in China's cities are the so-called urban villages – *chengzhongcun*. Urban villages were originally rural settlements located in suburban areas. Because of urban expansion, the agricultural land of these villages was gradually taken over for infrastructure and property development, and these villages physically became incorporated into urban built-up areas. However, owing to different population registration systems applied to urban and rural areas, these villages maintained their rural organisation and most farmers remained outside the formal urban management system. On account of less stringent planning restrictions and regulations, housing in these settlements became the prime locations for the poor, particularly rural migrants. Villagers built poor quality rooms one on top of another – some reaching five or six storeys – and rented them to migrants. Rural migrants also found the living environment more acceptable and less intimidating than in other settings because the landlords were farmers. Also, the rent was normally cheaper than in other areas.

Urban villages are common in all large cities. In the fast urbanising regions, such as the Pearl River Delta and the Yangtze River Delta, a large proportion of rural settlements ended up as urban villages. In Shenzhen, for example, several villages occupy a large proportion of land in the centre, right next to the central administrative and commercial districts. Original residents (farmers) no longer depend on agricultural activities for their livelihoods. Instead they engage in commercial or property-related business. In Dongguan, one of the fastest industrialising cities, factories were initially built around rural towns and villages and gradually merged together to form the city. In other cities, though the scale is smaller, hundreds of such villages can be found. For instance, in Xian city, within 190 square kilometres of the built up area, there are 187 such villages, and some of them are located in close proximity to the city centre.

Housing in urban villages located at the urban–rural interface zone is pre-dominately provided by individual families. Housing in well-established urban villages located deep inside a built-up area takes several forms: new houses built by individuals on their own plots for renting out, older houses left by original owners, and new housing or dormitories built by the village authority collectively for renting out. Most villages in Shenzhen city have reformed their traditional administration structure and have become so-called Shareholder Companies (SC) Ltd, in which the original villagers are shareholders. The collectively owned housing has been developed by the SC Ltd on collectively owned land and rented out. The rent from these properties is used to pay the SC Ltd management fees or for income distribution to the villagers. Each building owned by SC Ltd usually specialises in one form of housing; for instance, dormitories for single workers or units for families. Privately owned rental housing varies substantially between buildings and also within each building. Each private building reflects the wealth of the family that has built it. Richer families build better and taller buildings (plot size is similar for all villagers). Inside each building, the units available for rent also vary. Some are single-person units offering one bedroom while others are proper family units that comprise three bedrooms and a hall area.

Characteristics of migrant housing

This section discusses some common features of migrant housing such as tenure, quality, facilities, and physical and social environment. In inland cities, housing types for migrants are mainly simple single-storey shelters. Only a small proportion of housing takes the form of modern structures such as flats or self-constructed multi-storey private houses (Wang 2003, 2004). In coastal areas, most urban villages have been redeveloped and housing types are predominately multi-storey buildings. In both regions, however, the quality of the housing conditions is below the average standard of the city.

Migrants find their housing through two main sources: the private rental market or their employer. In 2004 Shenzhen municipal government conducted a survey of migrant housing which showed that 66.4 per cent of migrants had found their housing from the market and 20.5 percent had obtained their housing via their employers. In the same survey 9.6 percent of migrants were found to own their own housing while 2.5 percent reported living with relatives or friends (Shenzhen Municipal Housing System Reform Office 2004). This survey covered all kinds of migrants in Shenzhen including professionals and university graduates who did not have a local *hukou* registration.

Our study on migrant housing conditions in Shenzhen and other cities has focused on poor groups living in urban villages. The tenure distribution is presented in Table 6.1. Even though the surveys represented in Table 6.1 have been carried out in different areas and at different time periods, they show that most low income and poor migrants live in rental housing and that few own a home. In urban villages, some migrants may live in accommodation provided by their employers. On the whole this group differs from the construction workers and

Table 6.1 Migrant housing tenure in selected cities

Housing tenure	Shenyang	Chongqing	Shenzhen
Owners	8.7	1.2	1.4
Provided by employers	5.0	5.7	13.3
Rented from the market	82.6	81.7	83.6
Borrowed from friend/relative	1.9	2.5	1.7
Total	100.0	100.0	100.0

Notes: Shenyang and Chongqing surveys were conducted in 2000; Shenzhen survey in 2005/06.

the industrial production line workers who live on construction sites or in factory dormitories. Those who live in housing inside urban villages, provided by their employers, generally live in accommodation belonging to a small business owner. For instance, restaurant owners provide accommodation to their kitchen assistants, waitresses and waiters. Most migrants who live in accommodation provided by their employers are singletons and share with their colleagues. Our sample in Shenzhen shows that compared with single male migrants, a higher proportion of single female migrants live in housing provided by their employers (17.6 versus 11.4 per cent).

The main reason that migrants choose to live in urban villages is that rent is much cheaper than in properly built housing areas or estates. As a trade-off, living conditions in these areas are generally poor. Housing floor space in each unit is typically much less than in the new house units developed by commercial developers. It was reported recently that in 2007 average housing construction floor space in Chinese cities reached 27 square metres per person (http://www.cs.com.cn). Average housing floor space per person among migrants is far below this average in all cities. By way of comparison, in 2000 in Shenyang and Chongqing, average housing floor space was about 10 square metres per person and 23 square metres per household. In Shenzhen, average housing floor space among migrants varied. Of those who shared accommodation, floor space per person was less than 10 square metres. Of the families that rented a whole unit, the average floor space per person was around 20 square metres. Even in the same building, space used by migrants differed substantially. In one of the buildings that we studied, one family of three lived in a three-bedroom unit which included a small hall, a toilet, a kitchen and a balcony. They had some good furniture as well. Their babysitter lived with them. The couple ran a restaurant and were about to buy their own house. By contrast, in the unit next door, a two-bedroom unit with a hall was used by no less than eight people. A couple lived in one bedroom; four and sometime even five singletons used the other bedroom, and two singletons slept in the hall. There was no space for any other furniture and cooking and bathroom facilities were very poor.

Sharing a room or flat with other migrants is very common. Even though rents in urban villages are relatively low, few individuals can afford to rent a whole flat on their own. In Shenzhen a large proportion of migrants share with others (Table 6.2). Those who share rooms with other families or individuals have the lowest standard of living. On average four persons share a room and an individual has an average floor space of only 7.7 square meters. However in some instances over 20 people share a room, and some individuals have only 2 or 3 square meters of living space. Most halls in flats tend to be used as another bedroom for either children or other tenants. Although the majority of those who share a room are single, there are cases of two families (normally married couples, some with small children) sharing one room. If shared by married couples, wooden board, cardboard or curtains are usually used to keep some privacy.

Facilities inside migrant housing are generally poor. In inland cities, such as Chongqing and Shenyang, less than 15 per cent of migrants have exclusive use of

Table 6.2 Housing condition: sharing

House sharing	Chongqing and Shenyang	Shenzhen
Two or more individuals or families sharing one room	7.0	23.2
Whole family/individual live in one room only	50.9	16.2
One family use two rooms	28.5	0.6
One family exclusively use a whole unit	7.0	59.4
Other	6.6	0.6
Total	100.0	100.0

Notes: Shenyang and Chongqing surveys were conducted in 2000; Shenzhen survey in 2005/06.

a toilet, and over 90 per cent do not have access to a shower. In Shenyang, most of the migrant houses that we visited did not have proper heating systems for the very cold winters. In Shenzhen, 37 per cent of migrants did not have their own toilet, bathroom and kitchen; over 40 per cent did not have showers; and only one-quarter of them had air conditioning for the hot summer. Moreover, even though gas supply has become normal in major cities, most migrants still used coal or gas bottles as their main fuel.

Rooms and houses for renting to migrants are generally unfurnished. Migrants have to buy their own furniture. For this reason, the standard of furniture also varies across families. Household furniture often reflects the nature of the residents' work. Some migrants also use their rental home as a production place. Shop owners' houses, for example, look like shops or storerooms, and restaurant owners' houses resemble a kitchen. In Shenzhen, we visited a room of about 9 square metres shared by two married couples. Apart from beds, an old radio, two handwashing basins, a couple of stools and two sets of gas cookers, there were no other furnishings. Another couple with a child lived in the same area. They were waste collectors and their small room was filled with old newspapers and flattened cardboard. In one small room, three young men slept on the floor and there was no bed; they worked in shifts and the sleeping area was used in rotation. They did not have any furniture, just some cleaning utensils and their travelling suitcases. Household furniture provides a good indication of how long the tenants have been in the city and their long-term plans. Those who had stayed in the city for several years and planned to stay on had some reasonable furniture.

The external environment of the migrants' living areas is also poor (Yang 1996, 2000). In inland cities, most urban villages are not integrated into the planned built-up areas. Water and electricity supplies are basic and insecure. Sewage commonly runs through open ditches along the streets; roads are unpaved and become muddy during the rain; public toilets are at a distance from many houses and are smelly. In the coastal regions, building quality tends to be better, but still poorly designed. Buildings crowded together at high density cause poor lighting and ventilation. In Shenzhen and in other cities in the Pearl River Delta regions, high-rise buildings on small family land plots are so close to each other that the locals nicknamed them 'kissing buildings' or 'shake hand buildings'; neighbours can shake hands through their windows. In these villages, there is a dearth of green or open spaces.

High-density and poor-quality housing present public health and safety concerns. Most buildings used by migrant workers in inland cities are old and dangerous. Larger rooms are subdivided and separation materials are either woodchip board or cardboard. Electricity wires run everywhere. Coal fire stones or gas cookers are often placed inside the bedrooms. Owing to these problems, few migrants are satisfied with their living conditions. The single migrants we met who were sharing a room with others were particularly dissatisfied about their housing arrangements. Additionally the urban originated migrants we met were less satisfied than rural originated migrants, possibly because the latter tend to have lower expectations.

When we asked migrants to compare the housing conditions in the city where they worked with housing conditions back in their original home, a large proportion would say that their housing conditions were worse in the city. In Shenzhen where most migrants live in recently built housing, less than 30 per cent of migrants thought that their housing conditions were better than the conditions in their home town or village, while 56 per cent of migrants actually experienced a drop in their housing standard. Most migrants who we met had moved from place to place in the city. In inland cities, many of those moves were to be closer to their work in order to save time and money on travelling or to save rent. Not many moves were made to improve housing conditions. (Wu 2004, 2006)

Housing and living costs

Migrants, particularly poor ones, live in poor areas in cities (Ma and Xiang 1998; Mobrand 2006). Migrants do not have any assets in the cities, and rent is a major item of expenditure for them. Rent in the private sector is high in all Chinese cities. Most migrants work extremely hard to earn a living, but the average disposable income per person among migrants is much lower than the average for native city residents. At the same time, migrants' cost of living is much higher than that of urban residents who have a similar level of income. Comparison shows a distinct difference in the amount of income spent on housing by migrants and by other urban low-income families. Our study in Shenyang and Chongqing found that most poor urban residents spent less than 5 per cent of their income on housing while most migrants had to spend more than one-third of their income on rent. In 1999, the published monthly per capita disposable income of urban households was 447 *yuan* in Shenyang and 491 *yuan* in Chongqing. For migrants, if we include only the dependent family members living in the city, the monthly per capita disposable income for the same year was 363 *yuan* (405 in Shenyang and 307 in Chongqing). If we include family members (parents or children) back at home who require support from the migrant, the income per person falls substantially. Many migrants have between one and seven people to support. One of the main objectives of most migrants is to send money home each month. This reduces the disposable income among the family members living in the city. In 1999 in Shenyang and Chongqing, most migrants actually lived around or under the poverty line measured according to the World Bank benchmark of US$ 1 a day (Wang 2003).

The housing cost among migrants in Shenzhen shows a similar pattern to that for Shenyang and Chongqing. In 2006, 62 per cent of heads of migrant households earned less than 2,000 *yuan* per month, and among migrants' partners, over 76 percent earned less than 2,000 *yuan* per month (Table 6.3). The median income in the city in 2005 published by the municipal government was 2,164 *yuan* per month (Shenzhen Municipal Labour and Social Security Bureau 2005). The amount of rent paid by migrants is listed in Table 6.4. On average, migrants had to spend around one-quarter of their income on rent.

Table 6.3 Median and mean wages among migrants in Shenzhen 2006

	No. of respondents	*Median monthly income*	*Mean monthly income*
All head of households	745	2000	2749
HoH from urban areas	248	2900	3780
HoH from rural areas	497	1500	2235
Male	537	2000	3081
Female	208	1500	1891
Living alone	348	1900	2320
Living with family	397	2000	3125
All partners	222	1500	2082
Partner from urban areas	59	2000	3181
Partner from rural areas	163	1200	1684
Male	65	1500	2926
Female	157	1200	1733

Table 6.4 Rent paid by migrants in Shenzhen in 2006

	No. of respondents	*Average monthly rent*
Whole group	805	534
One person households	375	422
Male	217	403
Female	158	448
From urban area	153	512
From rural area	222	360
Multi-person households	430	632
Headed by a male	344	640
Headed by a female	86	600
Head came from urban area	118	741
Head came from rural area	312	590
All urban to urban migrants	271	612
All rural to urban migrants	534	495
All male headed household	561	548
All female headed household	244	501

Some observers may argue that many migrants are from poor rural areas and prefer poor-quality housing in order to save more money to send home. But for most of the migrants we met, poor quality housing was not a choice. With the income level and the cost of food, the only way for them to stay in the cities was to reduce their accommodation expenditure. Their wage was usually their only source of income and all expenditures needed to be covered by it. The composition of income among official urban residents is, however, much more complicated. The official salary is only part of the whole package. They often also receive various bonuses, subsidies, benefits and, for some, so-called 'grey income'.

Low income limits migrants' choices for housing. Other living costs are important as well. We asked migrants in the cities to rank their main expenditures. The cost of food always came first. In Shenyang and Chongqing, over 60 per cent of migrants ranked food costs as their largest item of household expenditure; in Shenzhen income in general was higher than in inland cities, and still, 42 per cent of migrants ranked food costs as their largest item of expenditure. Approximately 80 per cent of migrants in Shenzhen listed housing as one of their three top items of household expenditure. After food and housing, remittances were the next largest usage of migrants' wages (Lu and Song 2006). Not surprisingly little remained for other activities. In inland cities, most poor migrants do not have bank accounts and carry their small savings in cash with them all the time.

Managing migrant housing

Most municipal governments have not provided any large-scale housing for migrants. Commercial property developers have been mainly interested in the housing market for rich and middle-class families. As a result, migrant workers, particularly the poor ones, have been excluded from all new housing estates in cities (Wang and Murie 1999; Wang 2000) The rental housing offered by private landlords is normally located outside the control of the municipal government. Early government policies toward migrant workers were rather negative. Housing-related policies were concerned mainly with improving the overall urban environment, rather than with improving the environment in which migrants lived. Official reports often showed how migrant workers had created great pressure on urban infrastructure, exacerbated housing shortages, increased crime rates and caused employment pressure on the existing urban residents. Policies aimed to control the volume of migrants arriving in the cities, and to use an urban renewal process to clean up the poor-quality housing areas inhabited mainly by migrant workers. Many government departments were involved in the management of the migrant population and the most effective authorities tended to be those involved in policing, registration, taxation and family planning. Rather than providing necessary services to the migrants, these authorities often systematically discriminated or excluded migrants from areas of child education, employment and social welfare (unemployment, health and other benefits).

The unequal treatment of migrant workers has over the years created tensions between the migrant workers and original urban residents. In recent years, however, more and more Chinese researchers have begun to pay attention to the welfare of migrant workers themselves. They have praised the positive contributions that migrants have brought to the cities and highlighted the housing and living problems of migrants. Gradually, policy developers have themselves started to pay greater attention to the problems of migrant housing. Governments in several major cities including Shenzhen, Shanghai and Beijing, have carried out surveys on migrant housing conditions and demands. These studies have led to proposals to improve the living conditions of migrants. In the suburban town of Maluzheng in Shanghai, for example, the government has constructed migrant workers' housing estates. Small apartments and dormitories have been built and rented to major employers to accommodate their workers. The government in Dongguan, a new network city based on 32 towns, has actively encouraged large employers to build dormitories and rent them to industrial workers. Beijing, Changsha and Nanjing have all piloted housing schemes aimed at migrant workers (Shenzhen Municipal Housing System Reform Office 2004) but so far these efforts have remained largely at a pilot stage and the number of migrants to benefit directly has been small.

In 2004 the Shenzhen municipal government introduced the Rental Housing Comprehensive Management system which consists of several tiers of organisations from the municipal level down to the neighbourhood level. In the past the responsibility for migrant population registration and management fell to the police and public security authorities. Recently, the Rental Housing Comprehensive Management Offices (RHCMO) and local stations were set up throughout the city to help implement policies pertaining to the management of migrants. The RHCMOs have been given the power to

- identify and register private landlords and rental properties;
- identify and register tenants living at private rental accommodation (mainly migrants) and to copy information from their identity cards;
- record the migrant's employment situation;
- regulate and report safety and security related problems in rental housing (for instance, subdivisions created using wooden boards present a fire hazard and so need to be reported to the fire services and removed);
- record and manage health related problems;
- collect rental income tax from landlords;
- collect management fees from landlords;
- regulate and report illegal business and enterprises;
- monitor and enforce family planning policy among tenants.

As the name and the list indicate, the responsibilities of the RHCMO are comprehensive and wide-ranging, but its key attention has so far been directed at rental properties and tax rather than at the welfare of migrant workers. In 2006, the private residential property rental income tax rate was 8.22 percent.

This tax income goes to the municipal tax office. On top of the tax, there is also a management charge of 2 per cent for legally registered properties and 3 per cent for illegal buildings (buildings without planning permission). The management charges are used to run the RHCMO system. Both the tax and the charge have to be paid by the landlords, but these costs eventually fall on the tenants because some landlords simply pass them on to the tenants. In this sense the new management system has become a burden on migrants rather than a source of support for them. Nevertheless the migrants stand to benefit from building regulations pertaining to fire, health and safety. So far, however, the effectiveness of this system in improving migrant workers' living conditions has been limited by the desire of both landlords and tenants to avoid the tax and their reluctance to meet the RHCMO officials.

The registration of migrants by the RHCMO is less intimidating than the previous registration with the police, but even so, the police remain involved in the registration of migrants because the registration data is shared between the police and the RHCMO. When RHCMO officers visit a rental property, they record the migrant's personal information from their ID cards. This information is then copied into the record book in the office, which is in turn entered into a database. When the tenants move away, the record is removed. This registration provides the only data on migrant workers living in the city. Data on migrants published by the municipal government draws on this database. However, local officials indicated to us that the database is far from accurate. Not every migrant lives in properly regulated housing; not every landlord reports their tenants (for tax reasons); some migrants live in employer (factory owners) provided housing; and some factory workers live outside the factories' dormitories. No one knows how many migrants live in the city.

In 2006 in order to improve the housing condition of migrant workers, the central government issued an instruction to local authorities:

> Concerned government departments must monitor and manage the basic sanitary and safety conditions of living areas of migrant workers. Employers who hire large numbers of migrant workers could construct dormitories for them on permitted construction land. Economic and industrial development zones with large numbers of migrant workers can provide land for the construction of rental housing. Efforts have to be taken on planning, construction and monitoring of rural–urban interface areas with a concentration of migrant workers. Migrant workers' housing demands should be integrated into local housing development plans. If conditions permit, employers could arrange housing funds for migrant workers if they would like to purchase apartments in cities
>
> State Council, 2006

In the short term it is both unrealistic and unnecessary to accommodate all migrants in government subsidised housing. In our view, the gradual improvement of migrants' living conditions should be a long-term policy. While the effects of the policies for managing migrants and regulating their accommodation remains to be seen, some of the approaches adopted by local authorities may make the living condition of migrants worse. The poor conditions and the stigma associated with migrants' living areas have caused some urban authorities to plan large-scale redevelopment programmes. The main aim of these renewal projects is not to improve migrant housing, but to improve the general image of the city. Developers and municipal governments also see redevelopment as an opportunity to generate profit and income. These actions often displace poor migrants to other poor and peripheral locations. Most migrants living in Shenzhen's urban villages indicated to us that if the area is redeveloped, they will move to other urban villages, because this is where they can afford to live.

Conclusion

Over the past 25 years, migrants have made major contributions to the economic development of Chinese cities. The migrants occupy a range of niches in the urban economy ranging from construction and manufacturing to services and trade. The economic niche of the migrants tends in large part to determine their type of accommodation. Many migrants live on the premises of their place of employment, for instance in makeshift camps on construction sites or in crowded factory dormitories. Most others, particularly those working as traders or in the informal sector, live in run-down, marginal and peripheral areas in cities. Yet despite the variety in accommodation types, poor housing is a common aspect of most migrants' urban living experience. Indeed migrants' housing conditions are on the whole considerably worse than those enjoyed by the mainstream urban population.

Owing to migrants' 'temporary' residential status, limited opportunities for education, and lack of access to urban social networks, most are confined to poorly paid work. At the same time, market competition and the lack of proper legal employment protection enable urban employers to keep the migrant workers' pay as low as possible. Municipal governments often impose various taxes or charges on the facilities and services, including on the housing used by migrant workers. Moreover even though in many urban localities there are more migrants than official residents, the former remain excluded from basic social services such as child education, healthcare, housing subsidies and affordable accommodation which are provided by municipal governments and by other community organisations. Despite and perhaps because of this marginalisation, there are no meaningful organisations in the cities that represent the interests of migrants and there are no formal channels through which migrants can voice their concerns. Migrants are therefore excluded from the municipal policy-making process. Policies, local regulations and land use plans, which often directly impinge on migrants' living environment, are made without their participation. Hence, economic, social and institutional forms of exclusion reinforce each other in ways that mean for most migrants, there are few housing options and minimal prospects for affording decent accommodation.

Acknowledgements

This chapter draws on a project on Housing the Rural Migrants in China, supported by the British Academy. We would like to thank the excellent research assistance provided by Professor Jiansheng Wu and Miss Miaomiao Xie.

Notes

1 *Bangbang jun* – a special term in Chongqing – refers to labourers who helped others to carry goods in river ports, railway and bus stations and outside shops in the city. All of them carried a wood or bamboo bar (*bangbang* in Chinese) and a rope. They were extremely useful in a hilly city where other transport was difficult. It was reported in 2003, there were more than 10,000 *bangbangs* working in the city.

References

Beijing Municipal Bureau of Statistics (2003) *Beijing Statistical Yearbook 2003*. Beijing: China Statistics Press.

Davin, D. (1999) *Internal Migration in Contemporary China*. Hampshire: Macmillan Press.

Fan, C.C. (2001) 'Migration and labour market returns in urban China: results from a recent survey in Guangzhou', *Environment and Planning A*, 33(3): 479–508.

Goodkind, D. and West, L.A. (2002) 'China's floating population: Definitions, data and recent findings', *Urban Studies*, 39(12): 2237–2250.

Hussain, A. (2002) 'Final report on urban poverty in PRC', TAR: PRC 33448, Asian Development Bank. Available at: http://www.cs.com.cn/fc/02/200710/t20071010_1218930.htm

Knight, J. and Song, L. (1999) *The Rural-urban Divide: Economic Disparities and Interactions in China*. Oxford, New York: Oxford University Press.

Lu, Z. and Song, S. (2006) 'Rural-urban migration and wage determination: The case of Tianjin, China', *China Economic Review*, 17(3): 337–345.

Ma, L.J.C. and Xiang, B. (1998) 'Native place, migration, and emergence of migrant enclaves in Beijing', *China Quarterly*, 155: 546–581.

Mobrand, E. (2006) 'Politics of cityward migration: an overview of China in comparative perspective', *Habitat International*, 30(2): 261–274.

Murphy, R. (2002) 'Return migration, entrepreneurship, and state-sponsored urbanisation in the Jiangxi countryside', in Logan, J. (ed.) *The New Chinese City: Globalisation and Market Reform*. Oxford: Blackwell, pp. 229–244.

Shen, J. (1995) 'Rural development and rural to urban migration in China 1978–1990', *Geoforum*, 26(4): 395–409.

Shenzhen Municipal Housing System Reform Office (with Shenzhen Municipal Land Resources and Property Management Bureau and Project Group for Housing Problems of the Urban Temporary Population) (2004) 'Main report for the project on housing problems of the urban temporary residents' unpublished internal report, Mimeo, Shenzhen.

Shenzhen Municipal Labour and Social Security Bureau (2005) 'Indicative salaries for the Shenzhen City labour market in 2005', unpublished internal report, Mimeo, Shenzhen.

Solinger, D.J. (1999) *Contesting Citizenship in Urban China, Peasant Migrants, the State and the Logic of the Market*. Berkeley: University of California Press.

State Council (2006) 'The State Council Directives on Matters of Migrant Workers', State Council, Beijing, 28 March.

Wang, Y.P. (2000) 'Housing reform and its impacts on the urban Poor', *Housing Studies*, 15(6): 845–864.

Wang, Y.P. (2003) 'Living conditions of migrant in inland Chinese cities', *The Journal of Comparative Asian Development*, 2(1): 47–69.

Wang, Y.P. (2004) *Urban Poverty, Housing and Social Change in China*. London and New York: Routledge.

Wang, Y.P. and Hague, C. (1992) 'The planning and development of Xian since 1949', *Planning Perspective*, 7(1): 1–26.

Wang, Y.P. and Murie, A. (1999) *Housing Policy and Practice in China.* Hampshire: Macmillan and New York: St Martin's Press.

Wu, W. (2002) 'Temporary migrants in Shanghai: housing and settlement patterns', in Logan, J. (ed.) *The New Chinese City: Globalisation and Market Reform*. Oxford: Blackwell, pp. 212–226.

Wu, W. (2004) 'Sources of migrant housing disadvantage in urban China', *Environment and Planning A*, 36(7): 1285–1304.

Wu, W. (2006) 'Migrant intra-urban residential mobility in urban China', *Housing Studies*, 21(5): 745–765.

Xu, H. (2001) 'Commuting town workers: the case of Qinshan, China', *Habitat International*, 25(1): 35–47.

Yang, D., Park, A. and Wang, S. (2005) 'Migration and rural poverty in China', *Journal of Comparative Economics*, 33(4): 688–709.

Yang, X. (1996) 'Patterns of economic development and patterns of rural to urban migration in China', *European Journal of Population*, 12(3): 195–218.

Yang, X. (2000) 'Determinants of migration intentions in Hubei province, China: individual versus family migration', *Environment and Planning A*, 32(5): 769–787.

7 The making of a global dormitory labour regime

Labour protection and labour organizing of migrant women in South China

Pun Ngai

Introduction

Since the late 1970s China's economic reforms have brought about an unprece-dented surge in rural–urban migration. This continuous rural–urban migration flow was accompanied by the arrival of transnational corporations (TNCs) to China from all over the world, especially from Hong Kong SAR, Taiwan Province of China, Japan, the US and Western Europe. New generations of internal migrants have been working for transnational corporations (TNCs), which are either directly owned by or operate as joint ventures with large American and European companies, or their Chinese contractors and subcontractors in the export-processing zones of China's industrialized towns and cities.

The Fifth National Population Census of China (2000) estimated the number of internal migrant workers in cities at over 120 million. Economic migrants have moved from interior provinces such as Hunan, Hubei, Guizhou, Sichuan, Jiangxi and Anhui, to the southern and coastal provinces where the special economic zones (SEZs) are located. These migrant peasant-workers moved, either for short or long periods, from their registered place of residence but without a corresponding transfer of official household registration, or *hukou*. As a result, unlike permanent city residents, they have been long deprived of government-subsidized housing, education, job training, medical care and social welfare in the cities (Solinger 1999). Migrant workers, women in particular, are mostly concentrated in labour-intensive light manufacturing industries, such as apparel, electronics, shoes and toys, and the low-end service sectors.

The rise of China as a 'world factory' further signifies a new century of surplus labour drawn from rural China to fuel the global economy (Chan 2001; Lee 1998; Pun 2005a). Not only in coastal China, but more generally throughout the whole country, industrial towns and cities have been booming. With China's accession to the World Trade Organization (WTO), capital from the manufacturing industries, high-tech sectors and financial business continued to pour into China, provoking Western criticism that Chinese workers were increasingly stealing jobs from Western labour markets. However, there are growing concerns also among NGOs as well as academic circles over globalization, labour conditions and labour protection in post-socialist China. In spite of the increasing number of transnational codes of good conduct applicable to transnational companies, labour conditions in China remain generally precarious (Chan 2001; SACOM 2005). Globalization and price competition and the introduction of just-in-time production strategies by transnational corporations do not favour an improvement of labour relations in China (Pun 2005b). Instead, new dormitory labour regimes continue to be highly exploitative and generate huge hidden social costs for Chinese women workers in particular, while also giving rise to a social force that is silently resisting and challenging the existing social order.

In this chapter we first analyse the making of the new *dagong* class of internal migrant workers, and the emergence of a specific 'dormitory labour' regime for the new workers, predominantly women, in South China. The questions of concern are: Can this new working class be protected by the state through labour legislation or through the All-China Federation of Trade Unions (ACFTU)? Or can it be protected by transnational company codes of conduct? Besides these, are there other alternatives to empower workers and, if so, what are they? Drawing from a case study of a local NGO, the Chinese Working Women Network (CWWN), we provide a critical view of whether or not community-based labour organizing can bring about certain labour rights and worker empowerment in China.

Situated primarily in the SEZ of Shenzhen, CWWN struggles to survive together with migrant women workers who desperately seek labour protection mechanisms in urban China. Set up in 1996, CWWN is a non-profit NGO which aims to improve the lives of Chinese migrant women workers. Its objective is to fight for labour and gender rights, and promote grassroots empowerment and social justice in China. Due to the great difficulties associated with organizing migrant workers at the workplace level, CWWN bases itself in the migrant labour communities and attempts various projects to organize workers outside the traditional trade union model. Limitations, shortcomings and lack of genuine political space for women's participation make the achievement of these empowerment projects a constant challenge.

The making of a new *dagong* class

The acceleration of the global manufacturing production process in the wake of China's accession to the WTO has contributed to a dual process that underlies the making of a new Chinese working class. First, the global process has replaced China's old socialist pattern of industrial ownership and previous workforce composition, which has been constantly restructured since the mid-1990s. In 1981, the state-owned enterprises (SOEs) produced three-quarters (74.76 per cent) of China's GDP, while the collectively owned enterprises, which functioned as subsidiaries of state firms, generated another 24.62 per cent of GDP (Lee 2005: 4). The declining industrial importance of the SOEs became even more dramatic as economic reforms deepened in the 1990s. By 1996, 11,544 units of SOEs had gone bankrupt (Lee 2003: 74). The national importance of the state-owned and state-controlled firms in terms of total industrial output dwindled to only 18.05 per cent and 10.53 per cent, respectively, in 2001. Regarding the number of industrial employees in SOEs, the 1990 *China Statistical Yearbook* states that, in the same year, there were 43.64 million staff and workers, constituting 68.42 per cent of total national industrial employment. The neo-liberal ideology through the WTO's deepening involvement in market competition and corporate consolidation has contributed to massive lay-offs of state and collective workers in the new millennium. The former socialist employment systems have given way to the market forces of demand and supply. As privatization, mergers and bankruptcies changed the face of Chinese work units, or *danwei*, the number of industrial workers in SOEs numbered a mere 15.46 million by 2002, or only 41.46 per cent of the total industrial employment in China.[1]

The state-initiated transition to a market economy was accompanied by a sharp rise in the numbers of jobs in the private, foreign-owned and joint-venture enterprises that now dot China's coastal cities. A new working class of internal rural migrant labourers, or the *dagong* class (Pun 2005a), in contrast to the Maoist working class, has been taking shape in contemporary China. During the late 1970s and early 1980s, decollectivization released a massive labour surplus in rural areas. At the same time, the central government has facilitated an unprecedented surge in internal rural–urban migration by partially relaxing some of the restrictions of *hukou*, or the household registration system. Most transnational corporations and their Chinese subcontractors proved eager to recruit millions of peasant migrants in export-led industrialized zones. Until the early 1990s, the accepted number of 'floaters'[2] numbered about 70 million nationwide. By the early 2000s the number of internal migrant workers in cities had soared to 120 million, with some estimates ranging from 100 to 200 million (Lavely 2001: 3).

Women constitute a significant proportion of the rural migrant population in contemporary China. The development of SEZs across China, similar to the development of corresponding establishments in most other developing economies, was based on a massive harnessing of young workers, in particular of unmarried and recently married women (Gaetano and Jacka 2004; Pun 1999). By 2000, female migrant workers accounted for about 47.5 per cent of China's internal migrant workers (Liang and Ma 2004). In the manufacturing zones in coastal China, they made up about 65.6 per cent of all migrants.

Rural migrants have been identified as temporary residents who work in a city but lack a formal urban *hukou*, an urban registration status that confers on urban residents the entitlement to permanently reside in the city and obtain access to welfare and protection (Solinger 1999). The old but still-existing *hukou* system helps to create and maintain exploitative mechanisms of labour appropriation in Shenzhen and other Chinese cities. The maintenance of the distinction between permanent and temporary residents by the *hukou* system enables the state to shirk obligations to provide housing, job security and welfare to rural migrant workers. While China's economy needs the labour of the rural population, the survival of these labourers as people is of less concern. This newly formed working class, or *dagong* class, is prevented from laying down roots in the city. Worse, the *hukou* system, together with labour controls, is instrumental in perpetuating the ambiguous identity of rural migrant labourers in ways that obscure their exploitation. Does the state, and the society, regard these temporary labourers as workers or as peasants? The term *mingong* ('peasant workers' or temporary workers) blurs the distinction between peasant and worker (Pun 2005a; Solinger 1999) and preludes recognition of these people as labouring members of the urban working class. Thus, while local governments and foreign enterprises profit from the labour of migrant workers they can at the same time avoid any welfare-related burdens that would otherwise be due to the workers.

Labour control under the 'dormitory labour' regime

Because official and unofficial structures prohibit this newly formed *dagong* class from establishing its own community in urban areas, the onus for continuously securing a supply of labour is left to industry. This shift of responsibility from the state to the private sector creates China's 'dormitory labour regime', which contributes to the creation of a particularly exploitative employment system in the new international division of labour. As millions of migrant workers pour into industrial towns and cities, the provision of dormitories to accommodate these workers remains a *systemic* feature of global production processes. Irrespective of industry, location or the nature of capital, Chinese migrant workers – whether male or female, single or married – are accommodated in dormitories within, or close to, factory compounds in China. The term the 'dormitory labour regime' captures the emergence of these dormitory factories as the hybrid outgrowth of both global capitalism and the legacies of state socialism (Pun and Smith 2005). Importantly though, the dormitory labour system is not only a form of labour management, it is also a platform for labour solidarity, for labour resistance and the emergence of new labour relations.

What is noteworthy as China opens up to global production processes, starting with the Shenzhen Special Economic Zone in 1981, is the systematic provision of dormitories by employers for their work force. This has become the norm and has been extended to the majority of production workers. Moreover, dormitory accommodation in China fits neither the paternalistic mould identified in the West, nor the 'managerial familism' of Japan, nor yet the firm as a 'total institution' of the pre-reform Chinese state enterprise (Pun and Smith 2005). This is because contemporary China's dormitories are not intended to be long-term arrangements but to accommodate migrant workers for short-term employment; thus, the dormitory precludes a protracted relationship between the individual firm and the individual worker. Moreover, the Chinese dormitory labour system applies to companies irrespective of their production characteristics, seasonality, location or employer preferences. Most importantly, in China these dormitories are not provided by the enterprises so that the employers can secure labour loyalty or retain scarce skills. Rather, the dormitory labour regime serves mainly to secure the short-term availability of migrant labour and to maximize the use of this labour during the working day. In particular, companies aim to secure for themselves expendable youthful migrant labour, particularly female workers, thereby perpetuating an infrastructure that sustains China's precarious employment systems.

The factory workers' dormitories are attached to or close to a factory's enclosed compound. They are communal multi-storey buildings housing several hundred workers, with typically anywhere from eight to twenty workers per room. Washing and toilet facilities are communal and are located between rooms, floors or whole units, making living space intensely collective, with no area except that within the closed curtains of a worker's bunk offering some limited privacy. But these material conditions do not explain the role of the dormitory as a form of accommodation – as a living-at-work arrangement. Central to the dormitory form is a *political economy* that governs the grouping of typically single, young female workers. Separated from their family, home and their normal routine, these workers are concentrated within a workspace and subject to the prevailing rules and routine that totally subjugate their own personalities, and turn them into instruments of production. The subjugation experienced by the migrant workers is fully illustrated by the production control over the workforce, which operates according to the following seven strategies (Pun and Smith 2005).

1   The absolute lengthening of the workday: a return to an absolute, not relative surplus-value production.
2   The suppression of wage demands: a high labour turnover makes it more difficult for workers to engage in collective bargaining in general, and to demand wage increases in particular.
3   Easy access to labour during the workday: a just-in-time labour system for just-in-time production profits quick delivery order and distribution systems.
4   Daily labour reproduction: control of the reproduction of labour operates in the factory (accommodation, food, travel, social and leisure pursuits within a production unit).
5   Compression of the 'work life': ten years are figuratively compressed into five years owing to excessive working schedules and the production-based use of chiefly young workers.
6   Direct control over the labour process: limited formal consensual controls characterize workers' bargaining power, while a system of labour discipline imposes penalties on workers.
7   State and non-market interventions: external and internal state actions (e.g. *hukou* system) that restrict labour mobility and affect the overall labour process.

Labour rights and labour protection

These specific characteristics of China's dormitory production system and the wider exploitative labour regime of which they are examples, have undermined any pro-labour policies proposed by the central government. In recent years, the central government has introduced new regulations concerning, for example, minimum wage and working hours. The stated goal underlying these regulations is a unified legal framework for the protection of all workers against exploitation and inhumane treatment. On May 28 1993, the Standing Committee of the

Shenzhen People's Congress passed the 'Regulations on Labour Conditions in the Shenzhen Special Economic Zone' so as to institutionalize the labour recruitment system and govern labour relations at both the enterprise-level and the city-level.[3] The notable feature of these regulations is the important labour protection they offer, including minimum wage, working hours and social insurance for internal migrant workers throughout the city. However, precisely in the course of these government-initiated labour reforms, at least two deficiencies surfaced: first, labour policies and labour regulations are unevenly implemented at the local level and hence labour protection measures are seldom enforced; second, state and collective workers have been hard hit by economic restructuring, and the phenomenon further reveals that official statements on labour protection are often merely lip-service. Concurrently, as private and foreign capital flows increase to coastal cities and industrialized zones, pronounced competitiveness has taken hold among TNC factory suppliers and local enterprises, each trying to lower the cost of their just-in-time production, while trying to raise the quality of their products. This trend has become especially pronounced since China's accession to the WTO.

In view of the working and living conditions that characterize the rural migrant workers in South China, the Guangdong Federation of Trade Unions (GDFTU) issued an investigative report in 1994, stating that all 127 surveyed foreign-investment firms had violated national labour laws by imposing excessive working hours on their workers (Sun 2000: 179). The report's findings should not come as a surprise insofar as Taiwanese-invested enterprises wield a militaristic management style (Chan 2001). Corporal punishment, physical assaults, body searches and other unlawful acts are commonplace. Factories under South Korean ownership are also notorious for their harsh labour discipline and management practices; for example, managers beat female workers on the shop floor, force them to kneel down in public and similar acts (Chan 2001: 56–63).

When comparing the current working conditions with those of the 1990s, we found little improvement in the present situation. For instance, the past ten years have witnessed persistently low wage levels. Another report released by the GDFTU in January 2005 shows that the average monthly wage of 11.6 million peasant workers in Guangdong to be only 55 per cent of the average monthly wage of state and collective staff and workers. In other words, a majority (or 63.2 per cent) of the peasant workers earned between 501 and 1,000 *yuan* a month. Most alarming was the finding that, despite unprecedented economic growth, the overall 12-year increase of Guangdong's wage level for migrant workers amounted to a mere 68 *yuan*.[4]

Nevertheless, the minimum wage in Shenzhen used to be the highest among all of China's cities. The wage level of the Shenzhen SEZ between 1997 and 1998 was 420 *yuan*, compared to 315 *yuan* in Shanghai and 290 *yuan* in Beijing. Table 7.1 shows the Shenzhen government's annual adjustment of the minimum-wage standards for the period 2000–2006, taking into account inflation and the cost of living. Even with this meagre minimum wage compared to the rising living standards in the city, many foreign-invested or private companies still fail to

Table 7.1 Legal minimum wage in Shenzhen city, 2000–2006

Year	Shenzhen SEZ (yuan)	Outside the Shenzhen SEZ (yuan)
2000–2001	547	419
2001–2002	574	440
2002–2003	595	460
2003–2004	600	465
2004–2005	610	480
2005–2006*	690	580

Source: Shenzhen's Labour and Social Security Bureau.

* The effective date of the legal minimum wage level was July 1, 2005. In previous years, the adjustment was set on May 1, International Labour Day.

observe this basic legal labour regulation, and hence almost all the labour protection provided by the government are only lip-service.

Working conditions in the garment factories

Guangdong province in southern coastal China is the largest production base for Chinese garment exports.[5] According to the Chinese Ministry of Commerce, in 2004 Guangdong alone absorbed a record US$ 10.1 billion of the U$ 60.6 billion foreign direct investment (FDI) in China. Guangdong has seen a rapid population increase to over 110 million, making it one of the most populated provinces in a country of 1.3 billion people. Transient rural migrant workers make up more than 31 million of Guangdong's population.

The CWWN conducted research in ten small to medium-sized garment factory dormitories in Shenzhen in 2003–2004. Its reports on women workers' wages, working hours and occupational health reveal labour conditions that are punitive and exploitative. The workforce of each factory ranges from 50 to 200 people, over 70 per cent of whom are female workers responsible for sewing – young girls in their late teens, and middle-aged married women. The smaller factories are mainly owned by small subcontractors, who are mainland Chinese of Guangdong and neighbouring coastal provinces. The larger supplier factories for major international brands are funded by investors from Hong Kong SAR and Taiwan Province of China.

In terms of working hours, the Chinese Labour Law, in effect as of 1 January 1995, stipulates that a five-day working week should not exceed 40 hours and that overtime work must be limited to a maximum of 36 hours a month. However, almost all enterprises failed to observe these regulations, and an average a working day often lasts between 12 and 13 hours, for 6 to 7 days a week. When the production deadline approaches, management sometimes reduces lunch and rest breaks to a mere 30 minutes. To cope with the increasingly just-in-time production schedule, management often requires workers to work non-stop into the morning. In extreme cases, they are forced to work for 48 hours in a row. Total working hours in a week can thus add up to between 90 and 110 hours (SACOM 2005).

Of course, with demands like these, housing in factory dormitories serves an essential role by ensuring the round-the-clock availability of the labour force.

Under such work pressures, women workers suffer from a variety of occupational illnesses that include menstrual disorders, back pain, headaches, deteriorating eyesight, fatigue and respiratory problems. The situation is made worse by poor ventilation on the shop floor. Weaker female workers sometimes faint at their workstation, an occurrence that is especially common during the hot summer. Most of the employers provide no paid sick leave. Paid maternity leave, also required by law, is likewise neglected, although this is supposed to be a basic benefit.

The biggest problem cited by almost every female garment worker we interviewed was the illegal wage rate and their below-subsistence income. Between 1 May 2004 and 2005, the minimum monthly wage in the Shenzhen SEZ was 610 *yuan* (approximately US$ 73). Overtime pay is set at 150 per cent of the normal rate on regular working days, 200 per cent on weekends, and 300 per cent on statutory holidays. In terms of an hourly wage, this works out to 3.51 *yuan*. Overtime hourly payment on weekdays would be 5.30 *yuan*, and weekend overtime 7.01 *yuan* per hour. Thus, the average worker in small factories who has to work 13 hours a day all year round, and taking into account the specific piece rate of the factory which must be commensurate with the minimum wage set by the Shenzhen government, should earn 1,916 *yuan* for a 400-hour working month. However, workers earn only around 500 to 800 *yuan* a month. There are two main reasons for this: first, while most of the garment factories follow a piece-rate instead of a monthly wage system, the piece rate is illegally low. Second, the piece rate is never made transparent to workers. The wage slip declares a lump sum, which is not broken down into specific components. But fines and deductions, such as food and lodgings, are clearly spelled out and can amount to 150 to 200 *yuan* a month. Workers can barely afford the cost of living in big cities like Shenzhen.

In recent years, some of the better factories have invited the CWWN to conduct training and labour education programmes. In these factories, the skilled women workers can earn 1,000 *yuan* a month, but overtime work is still unlawfully imposed on these workers, and total working hours are very much the same as in smaller workshops. Given the combination of pitiful income levels with onerous living expenses, most of the women workers we interviewed complained about their below-subsistence income owing to the illegally low wage rates. This severe problem of underpayment has triggered increasing numbers of collective actions by production workers in South China since 2000.

Moreover, a majority of surveyed garment workers at small domestic private-owned workshops are not given labour contracts – a clear infringement of labour law in China. Without a legal working relation based on a labour contract, the workers are not covered under the government-regulated insurance scheme. If workers suffer work-related injuries they will have great difficulty in filing medical claims for medical treatment.

It is obvious that local governments cannot be counted on to stringently enforce labour laws. To promote economic development and attract foreign capital, local governments are deliberately lax in their supervision and, in return, receive generous taxes from company profits and bonuses from prosperous enterprises. Quite often, local officials are also investors in the bigger enterprises, and sometimes accept bribes from smaller factories to issue production safety certificates for example.

On the other hand, factory owners and managers argue that pressure from transnational clients is also to blame. A manager of a big garment factory in Dongguan in Guangdong province, described in the the *Financial Times* the pressure caused by social compliance rules and the practice of global 'lean' retailing:

> We are under enormous stress, customers place late orders, they change their orders part way through manufacturing and they pay their bills late. At the same time, they ask us to provide better training for our staff, better health and safety and better accommodation. We just cannot do it all.[6]

Corporate codes of conduct are yet to be fully implemented in major economic zones of the Pearl River Delta and Yangtze River Delta in southeastern China. Corruption, false statements and records and cover-ups are quite common (Pun 2005b). The Chinese manager of a large factory that supplies garments to multinational corporations in the Pearl River Delta economic zone in Guangdong, confessed to the *Financial Times* (April 21 2005) that the workers' time cards and salary statements were faked to meet the clients' codes of conduct. A team of six employees are assigned to prepare the forged documents, which are 'a perfect match' of the foreign buyers' requirements. This is merely one example among thousands. More importantly, this clearly points out the limits of factory audits. Workers are heavily penalized if they fail to give a model answer to factory auditors, and are financially rewarded for the 'correct' answer. For instance, when inspectors ask about working hours, workers are expected to respond that they work a standard eight-hour working day, with overtime of not more than three hours a day.

Lastly, an examination of collective bargaining power in the ten garment factories reveals that almost no trade unions operate in foreign-owned factories or private enterprises, in violation of Article 10 of the Trade Union Law (2001) that stipulates that any enterprise with 25 employees or more has to establish a *jiceng gonghui weijyuanhui*, or grassroots trade-union committee, under the auspices of the All-China Federation of Trade Unions (ACFTU), and despite the stipulation in Article 7 of the Labour Law (1995) that workers shall have the right to join and organize a trade union in accordance with the law. If an official trade union operates in a workplace, the union's function is often confined to entertainment and welfare activities, such as the organizing of balls and parties during festival periods. Our in-depth interviews with production workers revealed, in fact, that none of them had any idea about the organization and function of trade unions. Should there be strong disagreements concerning wages and overtime, workers either turn to their immediate shop floor supervisors for settlement of the issue, or quit the job. Collective bargaining through trade unions is unheard of among many production workers.

However, to portray garment workers as docile and passive is misleading. Accounts of labour unrest in official statistics of arbitrated labour disputes and other studies about strikes and protests show that migrant workers have increasingly resorted to collective action to fight for their legitimate rights. 'Workplace mobilization' in a single workplace rather than across factories is the predominant mode of organization.[7] There are subtle and varied ways in which women workers resist exploitation on a daily basis, particularly in the factory dormitories.

Attempts at labour organizing

In the context of this global dormitory labour regime, precarious employment conditions and infringement of labour rights are still prominent in China. The CWWN, a Hong Kong NGO, was set up in 1996 in the industrial zones of South China and started to organize women factory workers in Shenzhen, the first Special Economic Zone of China, with the aim of improving the lives of Chinese migrant workers. In so doing, it fights for the recognition and upholding of labour and gender rights, and promotes grassroots empowerment and social justice in China. Because of the great difficulties of organizing migrant workers at the workplace level, the CWWN is rooted in the migrant labour communities and promotes various projects to empower workers outside the traditional trade union model.

In its endeavour to secure a basis in mainland China, the CWWN not only approached migrant women workers but also various local department officials who were sympathetic to the plight of rural migrant women and willing to develop practical and feasible projects. Despite its belief in the self-empowerment of women, the CWWN nevertheless understood the constraints of organizing a viable space in China and sought to actively cooperate with various Chinese counterparts. The experience of the CWWN shows that working with official counterparts, though not the only possibility, led to the opening up of certain possibilities in this regard. For almost ten years, the CWWN has been able to carry out various grassroot empowerment projects in Guangdong province's manufacturing hub, the Pearl River Delta, concerning labour rights, occupational health and safety, gender equality and sustainable development. The CWWN also launched factory dormitory organizing initiatives and founded cooperatives for women workers[8] and, in order to provide empowerment programmes for Chinese women workers, conducts training workshops on labour and gender rights to enhance labour conditions in the workplaces. It also organizes cultural and educational activities to enrich the workers' social lives and encourages self-help solidarity. In addition to these continuous projects, the CWWN encourages workplace training, research exchanges and the sharing of experience among groups.

One of the unique programmes has been the CWWN's mobile van project, the Women Health Express (WHE), which runs through industrial towns close to the workers' dormitory areas, to reach out to migrant workers in distant industrial communities who could not personally come to its centre. An adapted 17-seat minibus, it contains a small medical clinic, library and cultural function centre with a TV, VCD and speakers for educational activities in open areas. This van functions as a mobile service centre to disseminate information on occupational health and safety, as well as to inform workers of and train them to assert their basic legal labour rights. The WHE began operating in the industrial areas of the Pearl River Delta on March 8 2000, and by the end of March 2002, the project had reached over 80,000 women workers. Despite this achievement, the project's shortcoming lies in its inability to operate stable workers' groups over a substantial period of time due to its mobile nature.

The CWWN is now seeking to speed up the integration of local women workers into its programmes so that active women workers can assume a more integral role in planning and organizing projects. Working with the Chinese migrant workers in South China in addition to the WHE, the CWWN also runs a women workers centre, a group for injured workers and a community centre for occupational health education.

The major problems and concerns of the women workers in the Pearl River Delta incorporate three main areas: (1) labour rights; (2) occupational health and safety, and (3) women's rights to independence and self-determination. These have become the major areas covered by the work of the CWWN.

For the daily operation of the projects, the CWWN established the following objectives.

1   To increase women migrant workers' awareness of community and occupational health and safety issues, especially regarding the prevention of occupational diseases and the protection of women's health.
2   To provide information and training on labour and employment rights, especially regulations concerning occupational health and safety and social insurance.
3   To offer basic health services such as simple physical examinations and occupational disease referrals.
4   To develop mutual aid and concern groups so as to reinforce migrants' awareness of labour rights, occupational health and safety and women's rights.
5   To enrich the social and cultural lives of women workers.

Community organizing

The Centre for Women Workers was established in 1996 to provide a platform for organizing Chinese migrant women workers in the SEZ of Shenzhen. Major organizing work includes labour rights education, protection against sexual discrimination in the workplace, sexual health education as well as training for migrants who return to their home areas. With the accumulated frontline experiences, the centre emphasizes the formation of self-organizing networks among migrant women workers in the factory dormitories, and develops workers' training programmes for migrant labour in its surrounding industrial communities.

The centre offers interactive programmes that are tailored to the learning needs of women workers. It trains volunteers to organize nearby dormitories and develop mutual support networks. Work initiatives include conducting small group discussions on labour rights, reading, handicraft, movies, photography, singing, drama and similar activities, thus providing women workers with a space to express themselves and hopefully articulate their collective identity, aspirations and needs as migrant women workers.

However, as so many women workers have to work long hours and cannot easily participate in the centre, the CWWN has had to reconsider its organizing work in order to reach the workers at the dormitories. Building a dormitory organizing network is therefore an urgent need in order to offer networking, training and other activities to women workers, and help them form mutual aid groups among themselves in their dormitories.

Organizational and educational work is now developed in the dormitory area, primarily among women workers, to promote awareness of labour rights, occupational health and safety, and feminist consciousness. The methods of organizing are quite diverse. Dormitory organizing activities are often carried out late in the evening around 11:00 p.m., so as to get in touch with the women workers who have to work overtime at night. Each time the organizers separate into small groups with one or two workers working at the visited factories, thus the workers are able to bring the organizers into their factory dormitories. In one factory workers usually form in different groups: those who have just come to Shenzhen; those who have worked there for half a year; those who enjoy particular recreational programmes, and so on. Different discussions and programmes are organized according to their needs. For workers who have just arrived in the city but who work in different factories, information on how to get used to working life in industrial towns and an introduction to labour rules and regulations, including minimum wages, are provided. Workers are also encouraged to visit each other and expand their social networks.

For those who already have some social relations in industrial towns, the organizers share experiences on how to improve working conditions. Over the years dormitory network groups have been developed to organize women workers and encourage them to communicate their legitimate demands concerning salaries and facilities to the management. In one factory, for example, 500 women workers succeeded in their demand for an increase in their basic salary and overtime pay, thereby improving working conditions by acting collectively. They also asked for a one-room-one-telephone plan after the factory agreed to install telephones. Furthermore, mini-libraries were established in factories and volunteers trained to assist in management.

Organizing of injured Chinese workers

Based on the WHE project, the CWWN trained a team of local organizers in China to help promote the self-organizing capacity of injured workers and to advocate for public awareness regarding occupational injuries and diseases in China. Localization is an important agenda for the implementation of this project. In the process, the CWWN also tried to network with local agents to help support the work of the mobile van. All these localized networks and staff form the important foundation for the future development of occupation health and safety rights. Intensive training and the transmission of knowledge on legal compensation according to Chinese law has been provided to a core group of injured workers since 2003. A Concerned Group for Injured Workers was formed by the workers themselves in November 2004. This group regularly visits hospitals in the industrial towns where injured workers are known to be. Most of these victims of work-related accidents had not been informed or educated about the potential hazards in their working environment, nor were they aware of their legal rights to protection and their entitlement to compensation in the event of injury. With the establishment of this group, it is hoped to ensure the empowerment of migrant workers and inform them of their basic labour rights, as well as to advocate for better occupational health and safety education and protection in China.

In order to further consolidate the injured workers group, the Occupational Health Education Centre (OHEC) was set up in September 2004. This centre strongly believes that education is the best way to prevent occupational diseases and industrial accidents. The frontline staff operates hotline consultation services, produces education kits, provides information on occupational health risk assessment, and organizes participatory training workshop for workers at plant level, with the goal of building a training and advocacy centre for occupational safety and health in China.

Services and activities include an OSH library, activity room, consultation, and developing a support group for fellow workers mainly for visiting patients and to train volunteers. The main training session subjects include information on the prevention of occupational health hazards, including poisoning, labour rights and interests, industrial injury and related rights and interests, the current situation of OSH in Asian countries and so on. Through such training, volunteers cultivate their capacity for spreading awareness about OSH as well as labour and industrial accidents and related rights and interests among their fellow workers, as well as their ability to participate in the daily management of OHEC.

In 2005, the centre took another small step forward by setting up a specialized unit of legal support for migrant workers. This unit provides migrant workers with systematic legal knowledge such as the protection of labour rights, compensation for occupational accidents and injury and social security as stipulated under the Chinese Labour Law. The staff compiles relevant policy materials and creates user-friendly training manuals. Audio-visual aids are also used for educational purposes.

Most of these community organizing or education projects have to rely on international or Hong Kong NGO experience to provide infrastructural support or training. The role of the state is minimal – the local Youth League appears to be the most active agent collaborating with NGOs on different projects, even though it has not been involved in actual implementation. Nevertheless, the involvement of the Youth League provides a semi-official opportunity for organizing activities, where women workers can take advantage of the possibility to train as active members in the empowerment process.

Workplace democracy?

The fundamental failure of the current monitoring system followed by most transnational corporations is that worker participation is mostly absent throughout the drafting and implementing stages of codes of conduct (Pun 2005b). In 2004, the CWWN, along with two partner organizations, created workers' committees in two factories to promote workplace democracy by building a worker-based monitoring model at the workplace level. The size of a workers' committee varies with the number of workers in a factory, but is generally between 12 and 14 persons; that is, a ratio of 1 committee member to 30 workers. The nomination and election process is an open and democratic one in which every worker casts a secret ballot during working hours.

The workers are trained in labour rights and means to monitor the application of codes of conduct at their workplace. The workers then conduct factory-wide elections to select their own representatives to the committee. The committee examines whether factory conditions comply with local labour laws and codes of conduct. The workers committee is conceived as an alternative mechanism to the top-down official trade unions governed by the All-China Federation of Trade Unions (ACFTU).

However, one of the biggest challenges for the CWWN has been to provide the workers' committee with continuing support. The CWWN trainers found that often workers are yet to be involved in key decision-making processes that directly affect them, such as working hours and overtime payment. Their negotiating power vis-à-vis the management on everyday production matters needs to be strengthened and the training curriculum updated to take account of their input. This attempt to promote workplace democracy seems still at its early stages and there is definitely a long way to go.

Conclusion

China has become integrated into the global economy, and is now a 'world workshop'. A huge pool of cheap migrant labour has been drawn upon to meet the needs of global production processes in the new international division of labour. Alongside China becoming the world's factory, migrant women workers sharing a collective identity are also forming an emerging new *dagong* force in Chinese society. The CWWN envisions crafting a civic space from the bottom up for the millions of migrant workers who form the backbone of the booming export economy, and it does so by training workers and local organizers and building social solidarity.

As countless workers are being drawn into this new private industry, old institutions and practices that allowed some measure of industrial democracy in the socialist context have proven ineffectual. New labour laws are not enforced.

In the new post-socialist China, there is little room for grassroots workers' voices or women's empowerment. In spite of new labour laws, official rhetoric about women's rights and the increase in transnational 'codes of conduct', exploitative labour relations sustained by the dormitory labour regime still prevail.

However, migrant women workers who risk abuse, work-related injury and death in the Pearl River Delta are beginning to empower themselves and invoke the rights granted to them on paper by the until now largely ornamental labour and gender equality laws of China. This trend may prove to be a step towards building a civil society and the expression of the people's voice in mainland China. The CWWN's work is aimed at just this target – empowering the people at grassroots level to create demand for workers' rights, particularly among migrant women workers.

Yet the development of this new rights-consciousness is being fostered against a backdrop of abuse and hidden costs created by the global dormitory labour regime supported by the hastily developed global factory chain. With the 'race-to-the-bottom' transnational production strategies adopted by TNCs, labour relations in China are difficult to improve without proper resources, mobilization and effective strategies. Because of the difficulties related to survival as a local NGO in China, the CWWN has spent a great deal of effort in ensuring its sustainability as an organization, which also affects its ability and power to organize labour.

All in all, the CWWN serves as an alternative community labour organizing model to fight for labour rights in the export-processing zones. It targets foreign investment and private companies which rely extensively on internal migrant workers, whose basic civil rights and labour rights are seriously violated. In addition to building labour networks through central organizing and the mobile van project, it also encourages cultural projects to help migrant workers fight collectively for their labour and feminist rights, and to strengthen workers' solidarity. A collection of oral stories of migrant women workers has also been compiled to foster common and collective working experience and labour consciousness.

Through supported structures for collective action, women workers can transcend differences in localities, ethnic origins, gender, age, work positions and the like, to achieve empowerment. The activities of the CWWN reported in this chapter are important in revealing that there exists in China a wide array of strategies and diverse forms of collective labour organizing which can be distinguished from the conventional trade union organizing model. Supporting such NGO initiatives is a valuable way for organizations concerned with labour rights more generally to help women migrant workers in China claim their legal entitlements.

Acknowledgements

The author would like to acknowledge Hong Kong Research Grant Council for supporting part of the field studies in this chapter, drawn from a research project on 'Living with global capitalism: Labour control and resistance through the dormitory labour system in China' (research period 2003–2005). I also would like to thank Jenny Chan Wai-ling for assisting in collecting research materials and conducting field studies. I am also grateful to all colleagues of the Chinese Working Women Network in sharing their valuable experiences in organizing women workers.

Notes

1 See *China Statistical Yearbook* (2003).
2 The migrant workers who tend to 'float' from one location to another have moved, either short term or long term, away from their registered place of residence and have done so without a corresponding transfer of hukou (the official household registration).
3 *Shenzhen Jingji Tequ Laowugong Tiaoli* (Regulations on Labour Conditions in the Shenzhen Special Economic Zone) (1993) defines *laowugong* (temporary hired labour) as those who work in Shenzhen without local permanent residential household status (Article 2). Residents in Hong Kong SAR, Macau SAR and Taiwan Province of China as well as foreign nationals working in the Shenzhen SEZ are not governed by the regulations (Article 53).
4 *Yang Cheng Wanbao,* quoted from *Apple Daily* 22 January 2005.
5 In addition to garment production, the share of leading industry groups in Guangdong include electronics and telecommunications, electric equipment and machinery, petroleum and chemicals, food and beverages, and building materials (*Guangdong Statistical Yearbook* 2004).
6 'Code of conduct implementation in china: laying a false trail', *Financial Times*, April 21 2005. Lean retailing' means the ordering of 'smaller and smaller quantities (...) and more often by the week instead of by the season (Ross 2004: 22). Giant retailers and brand corporations dictate the production, financing and shipment cycle for supplying manufacturers at the low end. Domestic and offshore factory workers are marginalized.
7 Sociologist Ching Kwan Lee (2003 and 2005), based on her ethnography, argued that migrant factory workers in South China tended to mobilize their resource at the workplace level during labour conflicts, for example, to get their back wages. Provided such collective actions are lawful, local governments do not suppress them.
8 The cooperative project set up in 2003 but failed to continue operation in 2005 due to inexperienced organizing and severe market competition.

References

Bonacich, E. and Appelbaum, R.P. (2000) *Behind the Label: Inequality in the Los Angeles Apparel Industry.* Berkeley: University of California Press.

Chan, A. (2001) *China's Workers Under Assault: The Exploitation of Labor in a Globalizing Economy.* New York: M.E. Sharpe.

Chan, A. (2003) 'A "Race to the bottom": Globalization and China's labour standards', *China Perspectives*, 46: 41–49.

Chan, K.W. and Li, Z. (1999) 'The *hukou* system and rural–urban migration in china: processes and changes', *The China Quarterly*, 160: 818–855.

Cheng, T. and Selden, M. (1994) 'The origins and social consequences of china's *hukou* system', *The China Quarterly*, 139: 644–668.

Gaetano, A.M. and Jacka, T. (eds) (2004) *On the Move: Women in Rural-to-Urban Migration in Contemporary China.* New York: Columbia University Press.

Lavely, W. (2001) 'First impressions of the 2000 census of China'. *Population and Development Review*, 27(4): 755–769.

Lee, C.K. (1998) *Gender and the South China Miracle: Two Worlds of Factory Women*. Berkeley: University of California Press.

Lee, C.K. (2003) 'Pathways of labour insurgency', in Perry, E.J. and Selden, M. (eds) *Chinese Society: Change, Conflict and Resistance*. London: Routledge Curzon, pp. 71–92.

Lee, C.K. (2005) 'Livelihood struggles and market reform: (Un)making Chinese labour after state socialism', occasional paper 2, United Nations Research Institute for Social Development. Available at: www.unrisd.org

Liang, Z. and Ma, Z. (2004) 'China's floating population: New evidence from the 2000 census', *Population and Development Review*, 30(3): 467–488.

Perry, E. (ed.) (1996) *Putting Class in Its Place: Worker Identities in East Asia*. Berkeley: Institute of East Asia Studies, University of California.

Pun, N. (1999) 'Becoming *dagongmei*: The politics of identity and difference in reform China', *The China Journal*, 42(1): 1–19.

Pun, N. (2005a) *Made in China: Women Factory Workers in a Global Workplace*. Durham and Hong Kong: Duke University Press and Hong Kong University Press.

Pun, N. (2005b) 'Global production, company codes of conduct, and labor conditions in China: a case study of two factories', *The China Journal*, 54: 101–113.

Pun, N. and Chan J.W. (2004) 'Community based labor organizing', *International Union Rights*, 11(4): 10–11.

Pun, N. and Smith, C. (2005) 'Putting transnational labour process in its place: dormitory labour regime in post-socialist China', paper presented at the 22nd International Labour Process Conference, 5–7 April 2004, Amsterdam.

SACOM (2005) 'Looking for Mickey Mouse's conscience: A survey on working conditions of Disney supplier factories in China'. Availalbe at: http://www.sacom.org.hk, accessed 10 November 2005.

Solinger, D.J. (1999) *Contesting Citizenship in Urban China*. Berkeley: University of California Press.

State Council (1993) *Shenzhen Jingji Tequ Laowugong Tiaoli* (Regulations on Labour Conditions in the Shenzhen Special Economic Zone), effective as of October 1, 1993.

State Council (1995) *Zhonghua Renmin Gongheguo Laodong Fa* (Labour Law of the People's Republic of China), promulgated on January 1, 1995.

State Council *Laodong Zhengyi Zhongcai Weiyuanhui Banan Guize* (Regulations on the Handling of Labour Disputes).

State Council (2001) *Zhonghua Renmin Gongheguo Gonghui Fa* (Trade Union Law of the People's Republic of China), amended and promulgated on October 27, 2001.

(2004) *Guangdong Tongji Nianjian* (Guangdong Statistical Yearbook).

(2004) *Shenzhen Tongji Nianjian* (Shenzhen Statistics Yearbook).

Zhongguo Laodong he Shehuibaozhang Nianjian (China Labour and Social Security Yearbook), 1994–2004.

Zhongguo Tongji Nianjian (China Statistical Yearbook), 1995–2003.

8 Civil society and migrants in China

Jude Howell

The Chinese economy has enjoyed unprecedented growth rates averaging around 10 per cent over the last three decades. By 2005 it had become the world's fourth largest economy, ousting the UK from this position. It is a significant trading partner of Europe, Japan and the US, has absorbed over US$ 60.3 billion in foreign direct investment[1] and has become a major consumer of the world's mineral resources. Its rapid growth has also contributed to a sharp decrease in the numbers of people living in abject poverty, official figures claiming a fall from 250 million to 26 million people between 1978 and 2004.[2] While China's economic miracle owes much to both state strategy and the gradual unleashing of market forces, it also rests on the industry and toil of millions of rural migrant workers who leave less lucrative rural areas in search of employment opportunities in the more prosperous urban heartlands. Without this army of relatively cheap labour that endures long hours and often arduous conditions of work, China's economic miracle could not have happened.

Compared to many other developing countries, the phenomenon of large-scale rural–urban migration is relatively recent, not least because of the tight controls over rural–urban mobility and urban residence in the pre-reform area. However the issues and social development challenges that migrant workers face are similar to those in many other societies, namely prejudice from urban-dwellers, insecure and often poor working conditions, restrictive government regulations, disparities in access to public services, and structural difficulties in organizing around common interests.

In this chapter we explore the role of foreign and domestic organizations in contributing to the enhanced welfare of domestic migrant workers within China. We begin by sketching the changing nature of civil society in China, which serves as a backdrop to organizing around migrant workers. In the second part we consider the particular challenges of working with migrants groups, such as raising awareness of their specific needs, countering urban prejudice, their often temporary location, obstructive national and local regulations and the broader constraints of organizing in China. We note also growing government concern for the plight of migrant workers, fuelled not least because of fears of social instability and potential regime threat. We then explore what kind of organizing has occurred in China, focusing on organizing on behalf of migrant workers and direct organizing by migrant workers. In the conclusion we consider the future prospects for organizing by and around migrant workers and the social and political implications thereof.[3]

Setting the scene: the development of autonomous organizing in China in the reform period

In post-liberation China the spaces for organizing independently of the Party state were sharply constrained. In order to restore social order after years of conflict, the Chinese Communist Party banned organizations it deemed threatening, co-opted others and instituted its own architecture of intermediary organizations.[4] Most prominent here were the mass organizations such as the All-China Women's Federation, the All-China Federation of Trade Unions, and the Communist Youth League, which served as the primary arteries linking the Communist Party to the defined constituencies of women and children, workers and youth respectively. These mass organizations served as key channels for the Chinese Communist Party to communicate downwards Party policy, and for grassroots views and perceptions to be reflected upwards, according to the principle of democratic socialism. Following a period of dormancy during the heady days of the Cultural Revolution, the mass organizations were gradually resurrected in the late 1970s.

With the onset of market reforms in 1978 under the leadership of Deng Xiaoping, the state began to loosen its hold over the economy, allowing market forces increasingly to determine resource allocation. The decollectivization of agriculture, the gradual development of rural markets and self-employed traders, the emergence of private domestic industry and the opening up of China's economy to global trade, all began to reshape the way the production of goods and services was organized. Aware that the partial withdrawal of the state from the direct management of the economy would require new ways of mediating market relations, the well-known economist Xue Muqiao put forward the idea of developing business associations.[5] These would serve as an intermediary link between the Party state and the market, taking on certain former state functions, enabling the Party to channel changes in policy downwards and business people to come together to address their needs. Trading, business and professional associations snowballed during the 1980s and formed the bulk of more independent forms of association during this period (Unger 1996; White *et al.* 1996; Wank 1995).

The growth of social organizations reached a peak during the tumultuous year of the democracy movement in 1989. Independent student organizations, autonomous trade union groups and democracy groups sprouted across the country, leading many China observers to interpret this wave of spontaneous associating as the rise of civil society. The tragic events of 4 June 1989, and the subsequent clampdown on independent organizing, halted the pace of growth of independent associations and led to disillusion among many China watchers about the prospects for civil society (Sullivan 1990; Huang 1993; Madsen 1993; Ding 1994). In the autumn of

1989 the State Council introduced new regulations governing the registration of social organizations. All social organizations were required to re-register under these regulations.[6] This allowed the Party-state to close organizations it was suspicious of, while at the same time encouraging the development of associations it deemed safe, reflecting the recognition that certain kinds of more independent associations were beneficial to the economic reform process. Over the next few years there was little growth in the number of registered social organizations as state officials proceeded cautiously and social organizations struggled to find government departments which would act as their supervisory unit (*guakao danwei*). China scholars became more cautious in their analyses, drawing back from the notion of civil society and highlighting the corporatist nature of many social organizations (White *et al.* 1996; Chan and Unger 1996).

As Deng Xiaoping regained hold of the political reins following his tour of the south in 1992, market reforms deepened and the political tension began to ease. In its efforts to counter international condemnation following the June 4 tragedy, the Chinese Communist Party was keen to host the 1995 Fourth World Conference on Women. This event catalysed the development of more independent women's organizations in China and indirectly fostered a more positive climate for autonomous associations (Howell 1997). With the opening of political space around gender matters, coupled with financial support from both the government and international donor and cultural agencies, women began to organize around social issues such as single motherhood, marriage law, professional and business interests, rural development and domestic violence. Apart from more formal types of organization, salons sprung up across China to discuss not only women's issues, but the concept of gender itself. Furthermore, the conference stimulated the development of women's studies courses and centres and established gender as a field of study (Du 2001: 178–179).[7] The Fourth World Conference on Women thus catalysed not only changes in the approach of the long-standing All-China Women's Federation to women's issues, but also in the landscape of independent organizing around gender issues (Hsiung *et al.* 2001).

Revisions of the registration regulations in October 1998 introduced more stringent requirements for social organizations.[8] Heightened government scrutiny of social organizations as part of this process led to the closure of some associations,[9] while some were unable to register as they could not meet the new conditions. Those seeking to establish new social organizations found the new regulations cumbersome and too demanding.

Despite the restraining regulatory environment the development of social organizations since the late 1990s has entered a new phase (Howell 2004). The two key features of this new phase are: (1) the rapid growth of organizations concerned with the social impact of economic reforms and (2) the tendency of new associations to bypass the restrictive regulatory framework. While professional, trades and business associations make up the bulk of registered organizations (Pei 1998), there has been a significant growth in associations concerned with welfare, poverty and inequality issues since the late 1990s. According to Pei (1998: 292) the number of national-level charitable groups and foundations increased from 2 in 1978 to only 16 in 1992, accounting for only 2 per cent of all registered national social organizations. Similarly, public affairs organizations, such as consumer societies, environmental groups and children's affairs groups, grew from zero in 1978 to only eight in 1992. Though exact figures are hard to come by, available information suggests that from the early 1990s onwards there has been a surge in the growth of charities, associations concerned with vulnerable and marginalized groups such as the sick, the rural poor, the elderly, wives of prisoners, and so on. For example, in 2001, China Brief produced a directory of 250 non-government organizations (NGOs) operating in the social development fields of health, education, environment, poverty, HIV/AIDS, gender and rural livelihoods, a substantial increase over 1992. While in the 1980s autonomous organizing reflected the interests of key beneficiaries of the reform process, namely entrepreneurs, professionals and researchers, from the late 1990s when some of the socio-economic consequences of rapid economic reform were becoming more evident, new forms of independent organizing began to address the concerns of those losing out from, or becoming marginalized by, market reforms.

The second feature of this new phase was the repertoire of bypassing techniques adopted by new groups. Given that the 1998 registration requirements were too stringent and that government units were often reluctant to take on responsibility for an association, new groups found other ways of pursuing their goals. These included affiliating to a larger social organization as a second, third or fourth layer organization, which required it only to submit a file rather than to register formally with the local branch of civil affairs. Some pursued their activities through participation in projects funded by donor agencies. Some groups expanded their portfolio of activities to take on new areas of work, thereby avoiding having to establish a new organization. Others carried on their activities as loose organizations with minimal resources. The Internet also provided a creative channel that could transcend physical boundaries and thereby the limits imposed by the government on organizing nationwide.[10] This circumvention of the registration regulations pointed not only to the ingenuity of citizens in finding ways to organize independently, but also to the more relaxed position of the government to certain forms of organizing. In particular, Party leaders were increasingly concerned about social stability, growing income and regional disparities, and the challenges of providing welfare, particularly in the context of state enterprise reform. To the extent that non-governmental groups were addressing social needs that could not be easily met by local governments, local leaders were willing for them to continue their work.

Since the 1998 regulations there has been no major change in the regulatory framework governing social organizations, though related regulations on foundations and public welfare donations were introduced in 1999 and 2004, respectively.[11] Although the idea of a law governing NGOs has been discussed extensively in policy and academic circles, it has yet to be passed at the National People's Congress. Hence the requirement that a social organization find a supervisory agency continues to be a key obstacle to the flourishing of the intermediary sphere. Nevertheless, at a meeting in the autumn of 2005, Party leaders actively encouraged the development of a charitable welfare sector in China. In this way the Party reaffirmed its commitment to a vision of civil society that is constituted ideally by charitable organizations providing welfare services, in line with the adage of 'small government, big society'.[12] In such a vision there is little room for organizations that challenge government policy or advance the interests of marginalized groups through advocacy or empowerment work.

It is against this background of civil society development in China that we have to understand the particular challenges of organizing around migrant workers and of groups involved in pursuing that aim.

Key challenges to organizing around migrant workers

Though rural residents have migrated in search of work in China's coastal cities for over two decades, organizing on behalf of migrant workers has only taken off during the 1990s. Moreover, any attempt by migrant workers to organize themselves in pursuit of their interests continues to be thwarted by central government. There are several social, institutional and political reasons why it has been particularly difficult to organize around the interests of migrant worker groups in China, compared to, say, the interests of entrepreneurs, traders or professionals.

First, migrant workers in China have encountered considerable social prejudice from urban residents. They are frequently viewed as the source of rising crime rates, as ill-mannered and poorly educated, attitudes that are often expressed in the ubiquitous phrase of migrant workers having 'low quality' (*suzhi di*). Given their low social status and negative image among urban residents and authorities, neither government officials nor the urban public were, until recently, willing to recognize that they had particular needs and interests. Up to the 16th Party Congress in 2002, central and local government leaders constructed migrant workers as a problem rather than as the subjects of social discrimination and the victims of often harsh, exploitative and unsafe working regimes or, alternatively, as entrepreneurial, risk-taking and dynamic economic agents.

Second, the social prejudice that migrant workers faced was exacerbated and perpetuated by a rigid institutional divide between urban and rural areas. With the stringent regulations governing rights to urban residence, migrant workers were seen as a 'temporary' population, whose needs and interests were not part of the institutional remit of urban authorities. As rural residents they had no rights to urban privileges and therefore no rights to make claims upon the urban authorities. Migrant workers thus became second-class citizens in their own nation state, denied equal access to housing, healthcare, welfare, social security and education in contrast to their urban counterparts (Solinger 1999). Local government leaders in urban areas framed migrant workers as a social group that needed to be restricted, regulated and controlled. As they perceived no obligation to provide such 'outsiders' to the cities with any entitlements, they did not foster an enabling environment for any non-governmental initiatives to organize services on behalf of migrant workers.

Third, the mobility of many migrant workers also made it difficult for any sustained migrant worker movement to develop. When migrant workers become dissatisfied with their working conditions, they are more likely to vote with their feet than to articulate their grievances or participate in collective action. Though migrant workers are staying on in cities and forming more permanent communities, mobility across workplaces weakens the possibilities of forming durable labour organizations and forging ties of workplace solidarity.

Fourth, as many labour activists lament, migrant workers identify more strongly around locality than class. In trade union experiments with direct elections of trade union cadres, sceptics have argued that it would be unwise to let migrant workers elect their own leaders as this would encourage 'clanism' and localism[13] (Howell 2006). However the rural roots of migrant workers are but one facet of the difficulties in developing a unified class consciousness and an identity across a fragmented working class in China. The other lies in the hierarchical divisions amongst workers, institutionalized by the Party-state,[14] whereby workers in state-owned enterprises perch at the top, whilst migrant workers are situated at the lowest end of the pecking order. Whilst the concept of *gongren* (worker) refers to workers in urban state and collective enterprises who, up until the mid-1990s when state enterprise reforms were accelerated, enjoyed relatively high social status, job permanency and various welfare benefits, terms such as *da gongmei, wailaigong* or *mingong* signal lower social status, an outsider, an interloper whose sweat and toil and mere presence unsettles the regular course of urban life. The discursive construction of the migrant worker as a separate category to the state-owned-enterprise worker, and one that counts for somewhat less in quality, entitlements and status, undermines the idea of a unified worker movement, based on a shared experience of the workplace and objective relation to capital.

Fifth, government concern about mounting social unrest, especially in rural areas, led to the introduction of restrictions on organizing by specific groups. In 1999 the Central Government General Office and State Council General Office issued a notice stating that special groups, such as retired soldiers, laid-off workers and migrant workers were prohibited from setting up organizations. This notice, in particular, has made it legally impossible for migrant workers to establish their own organizations.

Sixth, institutional territoriality has also been a barrier to migrant workers setting up their own associations. The case of the Migrant Workers Association in Zhejiang province in 2002 illustrates this point well. Though various local government bureaux, enterprise managers, village committees and migrant workers welcomed the establishment of a Migrant Workers Association, which indeed succeeded in registering locally with the Civil Affairs department, in the end it was the All-China Federation of Trade Unions (ACFTU) that put an end to this initiative. The ACFTU argued that the association was an attempt to set up a 'second' trade union in China, which was illegal.

A final reason why it is difficult to organize around migrant workers lies more generally in the lack of institutional channels for public influence on policy processes (Howell 2004). Over the last decade the Chinese government has tentatively begun to open up its legislative and policy processes to the public. However, this still remains quite limited, with most forms of public participation taking the form of consultation, which is mostly confined to individual elite experts. Whenever new central or local regulations on migrant workers are drafted, migrant workers are never consulted on their views, needs or interests. There is no advocacy structure or organization around migrant workers' issues. As will be seen in the next section, the majority of the organizations focus on service delivery rather than on any strategic attempt to influence policy change to the benefit of migrant workers.

Whilst there are limitations on organizing around the interests of migrant workers, there are also several favourable factors that augur more positive developments in the future. The first relates to the changing central government approach to migrant workers. Following the 16th Party Congress in 2002, central government leaders, such as Wen Jiabao, have paid considerable attention in their speeches to the problems facing migrant workers. This has led to several initiatives to increase the protection of migrant workers. For example, in 2002 the central government drew up a seven-year national plan to provide training for migrant workers.[15] In 2004 the central government introduced a three-year plan to pay outstanding wages to construction workers, which, by the end of 2003, amounted to over RMB 360 billion.[16] It has also initiated further discussion about amending the household registration system so as to make it easier for migrant workers to settle in cities. As a result, numerous cities have begun to ease restrictions on migrant workers gaining urban residency status and benefits. For example, at the nineteenth session of the 12th Beijing Municipal People's Congress, deputies voted to annul the ten-year old rules limiting migrants' access to housing and jobs.[17] In the south of China, factories in Guangdong have been facing a shortage of workers as migrants are attracted to cities elsewhere that not only offer higher wages but also promise urban residency.

Also indicative of a change in central government approach to migrant workers was the recognition of migrant schools that had developed spontaneously in migrant communities in response to the absence of state schooling and prohibitions on migrants attending local schools. Aware of the harsh working conditions under which many migrant workers labour, and the rising number of factory protests, central government leaders have also put pressure on the ACFTU to recognize the needs of and to actively recruit migrant workers.[18] However, as Wang Yuzhao, president of the China Foundation for Poverty Alleviation, highlighted in an article, such efforts hinge on their successful implementation and require national coordination (Wang 2005).[19]

A second favourable factor relates to growing government concern about social unrest arising out of increasing inequality, especially across urban/rural lines. On the one hand, this has not altered the government's position on prohibiting direct organization by migrant workers. On the other hand, it does create an environment where government agencies are seeking to address the symptoms, and sometimes causes, of these inequalities through policy change and are becoming more willing to tolerate non-governmental initiatives. The third and related factor is the government's growing recognition of the service-delivery dimension of civil society as an alternative to direct state welfare provision. Organizations that take on functions such as providing legal, job, health or housing advice to migrant workers should find it easier to get established and operate in the future. However, there will remain a fine line between service delivery and advocating the rights and interests of migrant workers that the government will not wish to see transcended.

In the next section we explore in greater detail the kind of organizing around migrant workers that has been possible so far in China. We examine both organizing on behalf of, and organizing directly by, migrant workers.

Organizing for and by migrant workers

It is difficult to track accurately the extent of organizing around migrant workers in China, as this can range from the more formalized activities of NGOs providing services for and/or acting on behalf of migrant workers, to the more fluid, loose and temporary collective action taken directly by migrant workers to articulate their grievances and demands. It can include non-governmental migrant workers' centres; migrant women's legal advice centres, and legal counselling centres taking up migrant workers' interests; the activities of individual lawyers in representing migrant worker interests; NGOs focusing wholly or in part on migrant workers, and donor projects and international NGO projects specifically targeted at migrant workers, usually working in conjunction with domestic NGOs. Collective action by migrant workers covers work stoppages, strikes, protests, petitions, letters of complaint and court cases.

This section draws upon available primary and secondary information, such as documentation of NGOs, donor reports, web-based materials, press clippings, statistical yearbooks and interviews with NGOs working on migrant worker issues carried out between 2000 and 2004.[20] Given the sensitivity of labour activism, the tendency to censor information about collective action, the lack of aggregated data on organizing by migrant workers, and the sheer size and diversity of China, this section does not pretend to sketch a complete picture of organizing around migrant workers. Instead, it provides an overview of the broad contours of organizing based upon available information.

Organizing on behalf of migrant workers

In this sub-section we examine the initiatives taken by NGOs, women's groups and lawyers, donor agencies and international NGOs to provide services and advice to migrant workers. It should be noted that many of these organizations will also be engaged in other projects and activities that are not directly aimed at migrant workers, but which nevertheless indirectly benefit rural migrants. It is beyond the scope of this chapter to cover this range of work; hence, the chapter concentrates on organization aimed directly at migrant workers.

As outlined in the previous section, organizing around the social effects of rapid economic reform, such as rising inequality, welfare provision and social justice, took off from the mid to late 1990s. Domestic NGOs concerned with the needs of migrant workers similarly started to emerge from the mid-1990s onwards in Beijing and Shenzhen. Here we describe some of the key initiatives in China. These take the form of migrant workers' clubs or centres, general and dedicated legal counselling centres, specific projects run by local NGOs and the activities of journalists and lawyers.

One of the first initiatives was the Migrant Women's Club in Beijing. This was set up in 1996 by the organizers of the publication Rural Women Knowing All, the first popular magazine to be aimed at rural women.[21] In September 2001, the magazine's organizer set up the Beijing Cultural Development Centre for Rural Women and a range of initiatives, including the Migrant Women's Club and the magazine, were brought under this organizational structure, while retaining their operational autonomy. The centre registered with the *Changping* branch of the Beijing Industrial and Commercial Administrative Bureau, thereby giving the organization legal status (http://www.nongjianv.org/english/school.htm, accessed February 2006). This process enabled the Migrant Women's Club to continue its activities without having to register with the local Civil Affairs Bureau or Industrial and Commercial Bureau, and could thus circumvent the difficulties of registration. The Migrant Women's Club aims to protect the legal rights of migrant women and create opportunities for their self-development. It is staffed by rural migrants and financed predominantly through grants from external donors. In recent years it has tried to increase financial support from domestic sources by holding charity dinners,[22] though support from international development agencies remains crucial for its activities. Over the past decade it has built up a portfolio of activities, which includes advice on legal rights, running classes in literary writing, legal awareness, gender awareness, psychological wellbeing and citizenship awareness, a magazine, *Mou Ning*, for migrant women, seminars on migrant women's rights, and an emergency relief fund for migrant women to support migrant women who have suffered injury at work, unfair dismissal or sexual assault (http://www.nongjianv.org/english/school.htm, accessed February 2006).[23] For example, the fund was used to support the court case of a female migrant worker who had suffered abuse as a domestic worker and, in another instance, to assist a female migrant in a legal battle with her employer after suffering work-related injury.

One of the most dynamic organizations working around migrant workers issues is the Chinese Working Women's Network. As Pun Ngai explains in the previous chapter, this was set up in the mid-1990s by feminists, labour activists, academics and social workers based in Hong Kong, who were concerned about the unsafe and exploitative working conditions facing many migrant workers. One of its first creations was the Shenzhen Centre for Female Migrant Workers, which was established in 1996 after the network successfully persuaded the local district trade union to grant it affiliation (interview, E37, 2001). Most of the paid staff and volunteers at the centre are former mainland migrant workers.[24] As described by Pun Ngai (this volume), the centre provides a space for migrant women workers to come together and organize activities, with a focus on labour law education, feminist rights, occupational health and safety, and long-term rural reintegration. The centre differs from other organizations working on behalf of migrant workers because rather than simply focusing on service provisioning, it places emphasis on the importance of an active empowerment approach to its work, which recognizes the gendered nature of power relations at the workplace (Pun and Yang 2004).[25]

In 2003, a second migrant workers' organization, known as the Culture and Communication Centre for Facilitators (*xiezuozhewenhua chuanbo zhongxin*) was set up in Beijing by two former volunteers of the Migrant Workers Club. The centre emphasizes the importance of the self-development of migrant workers and envisions its role as assisting migrant workers to address their own needs. In its own words, 'We are acting as a facilitator in workers' development in China with a view to making a common effort to facilitate and help people to help themselves' (http://www.facilitator.ngo.cn/english, accessed April 2006). The organization is therefore similar to the Chinese Women's Working Network in that its discourse reflects an empowerment agenda premised on an assumed relationship of equality with migrant workers. Its goal is not limited to providing services 'on behalf of' migrant workers, but is more transformative than reformist in that the migrants themselves are to become service providers (http://www.facilitator.ngo.cn/english, accessed April 2006). To this end the centre organizes a range of activities. It has established a Home for Workers targeting mainly female migrant workers and conducting such activities as legal awareness work, advice on occupational and psychological health, English tuition, entertainment and literature appreciation. The centre runs a counselling hotline, undertakes off-site visits to migrant schools and migrant dormitories, takes up legal cases, produces a newsletter for 'Friends of the Facilitators', runs seminars on issues such as occupational health and safety, and publishes a journal called *Dagong Shidai* (Migrant Workers Times) for migrant workers. The centre was initially financed by the founders and volunteers, but soon attracted numerous grants from international development agencies such as Oxfam Hong Kong, Ford Foundation China, Canadian International Development Agency and Action Aid.

Apart from these specific initiatives on behalf of migrant workers, non-governmental legal counselling centres play a key role in providing legal advice to migrant workers. Some specialized women's legal counselling centres include working with migrant workers in their mandate, such as the Beijing University Women's Legal Counselling Centre, the Shaanxi Women's Legal Counselling Centre or the Women and Children's Psychological and Legal Counselling Centre in Yunnan.[26] A minority focus specifically on women workers, the Fudan Women's Legal Counselling Centre in Shanghai being a case in point.[27] In addition to these general legal counselling centres, there are also centres that conduct work specifically on behalf of migrant workers. For example, one of the earliest non-governmental legal counselling centres focusing on migrant workers to be established is the Migrant Workers Document Handling Service Centre (*dagongzu wenshu chuli fuwubu*) in Panyu suburb of Guangzhou. Set up by Zeng Feiyang in 1998, the centre started out providing legal advice to migrant workers on issues such as compensation for work injuries and unpaid wages (interviews, October and November, Hong Kong, 2002). Within four years it had handled 540 individual cases, most of them dealing with claims for industrial accident compensation, unpaid wages and unfair dismissals. It soon developed its activities further to include outreach work for injured workers and outpatient departments, training concerning legal rights for workers, the publication of leaflets, a monthly journal, *gongyou* (Worker's Friend), and the publication of a book 'I am a Migrant Worker' (*wo shi dagongzhe*), relating the lives of migrant workers in the Pearl River Delta. For the first four years, the centre relied upon voluntary labour and contributions from local donors, but has since attracted external sponsorship. Unable to find a sponsor, the centre could not register as a social organization under the Department of Civil Affairs and registered instead under the local Department of Industry and Commerce as a private company (China Development Brief 2002).

Similarly, the Centre for the Protection of Migrant Workers' Rights (*wailai wugong renyuan quanyi baozhang zhongxin*) in Shanghai was set up in February 2001 to provide legal advice to migrant workers and to support them in adjusting to urban life. Its activities include a telephone hotline, the production of a handbook on the legal rights of migrant workers, outreach work, and promoting media coverage of migrant workers' issues (China Brief 2001: 234). In Hainan province, the Law Research Institute of Hainan University and the Migrant Workers Management Association established a legal clinic in 2004 to advise vulnerable groups such as migrant workers, and received an EU grant for one year in 2004 to support its activities.

As well as specific migrant workers' centres and legal counselling centres, NGOs engaged in corporate social responsibility activities have also targeted migrant workers. The first and most prominent such NGO is the Institute of Contemporary Observation (ICO) (*shenzhen dangdai shehui guancha yanjiu suo*), set up in Shenzhen in March 2001 by Liu Kaiming, a former journalist specializing in labour issues, with the support of Oxfam Hong Kong. This seeks to promote the development of workers and corporate social responsibility (CSR) through research, rights advocacy, advice on CSR, training and legal assistance (www.ico-china.org, accessed April 2006). Together with University of California at Berkeley, it established in March 2004 a Migrant Workers Community College. This provides cultural, rights and professional training in factories and in the community for migrant workers. Courses cover English, computer skills, occupational health, law and AIDS prevention (Tang 2004). It also received support from the Ford Foundation in March 2003 for the project China Labour Research and Support Network, which brings together researchers, individuals and service organizations and provides a platform for information exchange and capacity building. Compared to many NGOs, especially those working on labour issues, the ICO is relatively well endowed. Apart from funding from foundations, development NGOs and universities, it also generates income through conducting social responsibility audits for international buyers. The ICO's financial success lies in its strategy of working with different stakeholders such as companies, workers and government. In an interview with a reporter, Liu Kaiming emphasized that his approach is not to confront government or encourage workers to form independent workers' unions, thus, publicly at least, eschewing any agenda of empowerment (Tang 2004).

Apart from NGOs and donors, individual lawyers and journalists have played a key role in raising public awareness of the inequities and discrimination experienced by migrant workers. The journalist Jin Yan, for example, wrote a newspaper article in the official daily of Harbin city, northeast China, exposing companies for failing to pay migrant workers. The publicity given to these companies led to migrant workers' finally receiving their pay. The renowned labour lawyer Zhou Litai has boldly exposed the plight of workers injured and maimed in industrial accidents and fought through the law courts for compensation.

While all these organizations have been initiated and driven by Chinese intellectuals sympathetic to the plight of migrant workers, there are also some organizations which have been inspired by local government officials to address concerns over social stability. A key case in point is the Haikou Migrant Workers Management Association. This was founded in 2003 and built upon a swathe of various government initiatives to support migrant workers. Unique among these were the Homes for Migrant Workers (*wailaigong zhi jia*), the first of which was set up in 1996 and heralded as the first of its kind in China. By 2005, Haikou city had established five such Homes for Migrant Workers, providing advice on family planning, health, employment, legal rights, services for women and children and for children's education. The Migrant Workers' Homes were a government initiative to address the perceived problem of social order brought about by the rapid influx of migrant workers. In 2003, the local government inspired the establishment of the association, not least so that it could decentralize certain governmental functions, such as running the Homes for Migrant Workers. In this way it could delegate them to the market and social organizations, and divest itself of its financial contributions to the association (Gu 2005). In July 2004, the association registered as a social organization falling under the supervision of Longhua District Social Order Comprehensive Management Committee (http://www.wailaigong.net). Though the association became nominally independent, it had the character of a government-organized NGO, as reflected in its functions, structures and its resource base.[28] This separation from government has created financial difficulties for the association, though a one-year grant from the EU in 2005 provided some temporary relief (http://.law.hainu.edu.cn/flzz/jianjie.htm, accessed February 2006).

Similarly, the Shenzhen Migrant Workers Association (*Shenzhen shi wailaigong xiehui*) resembles in its structure, functions and language a government-organized NGO. This was set up in March 2004 in the Bao'an District of Shenzhen City as a non-profit people's organization to protect the rights of migrant workers. In May 2005 it became affiliated to the Shenzhen Chunfeng Labour Dispute Advice Service Bureau. Its committee structure encompasses a range of government functions, such as legal protection, culture and education, labour and capital mediation, economic development and cooperation, and fund-raising and management. The founder of the organization was formerly a trade union official in Hunan who migrated to Shenzhen in search of work. In 2003 he gave up his work to start the association. The main activities of the association, as described on the website, are legal advice to migrant workers, a hotline, assisting the government to strengthen the legal awareness of migrant workers, organizing cultural events, helping workers in difficulty and conducting research (http://www.szwlg.com, accessed February 2006). Both the language of the website and the nature of some of the activities, such as organizing entertainment and supporting workers in difficulty, echo some of the roles played by the ACFTU. Unlike the Shenzhen Centre operated by the Chinese Working Women's Network and the Facilitators' Centre in Beijing, neither the Hainan nor the Shenzhen associations demonstrate any substantive or discursive agenda to empower workers.

From interviews, website and documentary information it is apparent that grassroots organizations working on behalf of migrant workers have relied considerably on donor support to finance their activities. From a donor perspective rural migrants constitute a potentially vulnerable group facing inequality and social exclusion. Donor programmes and projects targeting migrant workers tend to revolve around the areas of health, education, and legal rights and interests. Some of the key players here are the UK Department for International Development (DfID), the EU, the Canadian International Development Agency (CIDA) and Asia Development Bank. The following examples illustrate the range of donor support to rural migrants.

EU support to migrant workers focuses on their legal rights. In 2004, for example, it granted EUR 60,000 (US$ 77,600) to the Migrant Worker Management Association of Longhua District, Haikou, to protect the rights of migrant workers, and EUR 55,518 (US$ 71,800) to the Centre for Women's Law Studies and Legal Services at Beijing University to provide legal aid on women's labour rights. Similarly, CIDA provides grant support under its Civil Society Programme for the establishment of non-governmental legal aid offices.

Apart from focusing on legal rights, some donor agencies have been supporting employment initiatives for migrant workers. The UK DfID and the Government of Japan co-financed an employment creation programme set up by the International Labour Organization and the Ministry of Labour and Social Security in 2004, where the second phase focused specifically on rural migrant labourers. The aim of the 'Start and Improve Your Own Business' programme was to provide comprehensive support to rural migrants to set up their own businesses. This takes place against a background where the government is encouraging the permanent integration of migrants into urban areas and their inclusion into government-run and employment promotion programmes (Perrement 2005).

Basing its approach on social justice and gender equity, the UK DfID China Office has sought to develop support mechanisms through project initiatives to protect migrant workers in the process of migration and rapid socio-economic change.[29] According to an evaluation carried out in March 2005 (Murison 2005: 8), DfID China's approach to supporting migrant workers is distinct from other DfID country offices in its emphasis on gender equity and its efforts to mainstream gender into donor and government policies and practices. For example, DfID's £3 million project (2004–2008) to prevent trafficking in young girls specifically indicates how to mainstream gender into its processes by, among other things, disaggregating data according to sex, and consulting girls and families on their needs. Similarly, its project on support to poor rural girl adolescents seeks to empower young girls to make informed choices about their future, gaining life and livelihood skills in the process. DfID's approach to migrant workers dovetails well with the Chinese government's Eleventh Five-Year Plan, which seeks to address the issues of social security in rural areas and the inequalities of access faced by migrant workers in urban centres.

Many international NGOs and foundations also take up the needs of migrant workers through specific projects. Some of the main foreign NGOs are Oxfam Hong Kong, Asia Foundation, and Kadoorie Foundation. Oxfam Hong Kong, for example, has provided grant support to a range of organizations and initiatives concerned with various issues facing migrant workers, such as legal rights, children's education and health. This has included over a decade of support to a service centre for migrant women workers in Shenzhen; support to a service centre for migrant workers from Inner Mongolia in Beijing that provides occupational training and legal advice to workers in Beijing; a grant for a conference organized by Rural Women Knowing All on the impact of the household registration system on migrant women workers, and support for industrial accident victims in Shenzhen so that they can pursue compensation claims through the courts, as well as research into advancing the legal rights of industrial accident victims. In 2004–2005, for example, Oxfam Hong Kong supported research on occupational diseases for migrant workers in Guizhou, the provision of legal services and training for migrant workers in Guangdong, Beijing, Hunan, the establishment of a Migrant Worker Labour Centre in Qingdao, and research on the education of migrant workers' children.

The Asia Foundation has supported a range of activities to provide services to migrant women workers and promote policy reform dialogues on the rights of women workers in China as part of its broader remit to improve the social, economic and political rights of women in Asia. In particular, it has awarded grants to three separate projects providing health, educational and counselling services to migrant women workers in the Pearl River Delta. On the basis of these projects, the Asia Foundation has launched a nationwide initiative on worker protection and labour rights education. In conjunction with a foreign retail company it set up a new scholarship scheme for female migrant workers to access education and vocational training, providing an opportunity to enter such fields as tourism and management.[30]

The Kadoorie Foundation has supported a Community Support Network for migrant workers in Guangzhou who have suffered injuries in industrial accidents. This network assists workers to access rehabilitation services and vocational retraining and also carries out occupational health and safety training in workplaces. The Amity Foundation has supported a number of educational projects to improve the conditions in the schools of migrant workers' children and to provide leadership training for the heads of these schools. Action Aid has similarly given support to migrant schools that are often set up by migrants themselves for their children.

With increasing donor support to HIV/AIDS work in China since 2000, and a more open and determined effort by the Chinese government to address this health issue, non-governmental activity in the field of HIV/AIDS has spiralled. Numerous NGOs and bilateral and multilateral donors have targeted migrant workers in HIV/AIDS awareness programmes. The Swedish NGO RSFU (the Swedish Association for Sexual Education), for example, started a project in 2001 to increase awareness of HIV/AIDS and condom use among migrant workers in Shanghai. This project has involved a baseline survey of factory workers, the development of a training manual, a health education newsletter, posters and a mobile exhibition. Similarly, Marie Stopes China, established in 2000, has targeted youth and migrant populations in its programmes on sexual healthcare and education, with AIDS prevention work cutting across all of its activities. The World Bank has provided support to the family planning associations in Guangxi and Fujian to carry out AIDS/STI education and condom promotion among migrant workers. World Vision is similarly developing an education tool-kit on HIV/AIDS for migrant workers. Further examples of international NGO involvement in sexual health education and outreach for migrant workers can be found in the chapter by Caroline Hoy in this volume.

There are also consultancy companies that work on corporate social responsibility issues and whose work bears directly on migrant workers. A prime example here is Impactt, a UK-based consultancy/NGO, which established the Impactt Overtime Project in 2001. This project brought together retail companies such as Debenhams, Hennes & Mauritz, Kingfisher and Body Shop International to work with Chinese organizations and supplier factories to identify sustainable ways of reducing excessive overtime work. The project emphasizes the importance of improving efficiency and human resource management, of change over time rather than immediate compliance, and raising productivity while reducing working hours.

A key concern that emerges from this overview of organizing on behalf of migrant workers is the dependence of many of these organizations on donor funding. While donor support plays an important enabling role, it also raises issues of sustainability if donors withdraw from China. The UK DfID, for example, is planning to substantially reduce its aid to China as of 2010, whilst the Japan International Co-Operation Agency and some bilateral organizations have already closed their programmes in China. The challenge for such organizations will then be how to raise money for their activities in a context where alternative sources of funding, such as from governments, enterprises or the general public are not readily available. This relates in part to NGOs' lack of familiarity with fundraising techniques in China – though recent capacity-building initiatives by donor agencies are seeking to address this – and also to the relatively recent presence of NGOs in China and a lack of familiarity among the general public with, and confidence in, their work.

Organizing by migrant workers

While the number of organizations working on behalf of migrant workers is growing, any attempts by migrant workers to organize themselves have been short-lived, even where local governments have supported the idea of migrant workers organizing themselves. Two cases illustrate this point well, namely the Migrant Workers Association in Ruian, Zhejiang province, in 2002, and the Migrant Workers Management Association in Longhua District, Haikou city, Hainan province.

In response to ongoing problems with industrial relations between migrant workers and employers in Ruian, Zhejiang province, migrant workers, with the support of the local authorities, set up a migrant workers association in 2002. Relevant government departments, such as the Civil Affairs Bureau, Public Security and the Labour and Social Security Bureau, all viewed the association as a pragmatic and effective solution to the ongoing problems. It was formally established in April 2002. The success of the association was reported in local papers, but soon caught the attention of higher levels of the All-China Federation of Trade Unions (ACFTU). The ACFTU was concerned that the migrant workers' association was a 'second trade union' and therefore illegal (Xie 2002; Pan 2002). As a result, the Migrant Workers Association, which had proved effective in calming troubled industrial relations, was prohibited in July that year.

In the case of the Hainan Migrant Workers Management Association, as discussed in the previous section, the local government was seeking ways to hand over responsibility for running the five 'Homes for Migrant Workers' to non-governmental agencies. This was in part because it was difficult to coordinate the running of five separate migrant workers' homes, and also because it did not reach all the workers in the city (Gu 2005). So, the government sought to create a structure that was broader in scope and able to reach out to all migrant workers. In 2002, Longhua District Politics and Law Committee decided to set up an association for migrant workers, namely the Migrant Workers Association, to enable migrant workers to organize themselves and provide a bridge between the local government and migrant workers. At the end of 2002 the small group charged with preparations for the Migrant Workers Association submitted a document to Haikou City Civil Affairs Department. However, as the central government regulations prohibited certain groups, such as migrant workers, laid-off workers and retired soldiers from forming organizations, the district government was required in October 2003 to submit another document proposing a change of name and constitution. The new association, now renamed the Migrant Workers Management Association, was now officially established. Though it claims to be the first association organized by migrant workers,[31] in fact the impetus to form the association came from local government. It was, therefore, a top-down initiative that recognized the need for 'the subjects' of organizations to have a more direct role in organizing around their needs. As such, it was also part of a process of renegotiating the boundaries between government and civil society.

While most organizations working on behalf of migrant workers address issues of legal rights, health and education, there are also a few that pursue these goals through cultural means. The Young Rural Migrants Arts Troupe was set up in May 2002 to perform music, dance, comedy and *kuaibanr* recitations for migrant workers on construction sites, in factories and in the communities where they live. The troupe members are themselves migrants, often holding full-time jobs as cooks and security guards. In late 2002, Sun Heng, along with other troupe members, registered the Home of Rural Migrants (*nongyou zhijia*) Culture and Development Centre in the north of Beijing as a company under the Bureau of Industry and Commerce. Apart from the performing arts, the centre also runs computer training courses, a mobile library and legal counselling services. In 2003 it received a two-year grant from Oxfam Hong Kong, which enabled it to pay full-time salaries for three troupe members (Qian 2005).

Migrant workers have also organized spontaneously around the education of their children. Unable to pay the high costs of primary and secondary schooling in the cities, migrant workers have set up their own community schools, employing teachers from the rural areas. However, as is discussed in the chapter by T.E. Woronov in this volume, local authorities have often viewed these schools as illegal and as failing to provide standard education to the children. In Fengtai District, Beijing, for example, local authorities closed down over 50 migrant schools (www.actionaid.org/china, accessed March 2006).

Given the severe difficulties migrant workers face in establishing their own organizations, collective action tends to remain in the form of spontaneous protests, strikes and work stoppages at the workplace. While it is relatively easy to collate general information about donor projects supporting migrant workers, the work of NGOs, or the number of disputes taken to arbitration committees, information about the number of labour protests and the different forms these take is more difficult to compile. Labour activism in China, such as strikes, protests, sit-ins, demonstrations, petitions, publication of leaflets and manifestos and the setting up of independent organizations is highly sensitive, as it challenges a key ideological premise of the Chinese Communist Party; namely, that the Party represents the workers and hence that workers' and Party interests coincide. Information about workers' protests is piecemeal and subject to censorship. Organizations such as the Hong Kong-based NGO China Labour Bulletin, headed by the labour dissident Han Dongfang, and Human Rights Watch, play a crucial role in tracking labour protests and publicizing disputes. There are official annual statistics available in the yearbooks of the ACFTU on the number of labour disputes that have been settled through arbitration committees. There is also a wealth of information in legal counselling centres and among lawyers about specific cases filed by workers against their employers, though there is no system for drawing this information together.

According to available information, there were over 58,000 mass protests in China in 2003 (Ming Pao website, Hong Kong, 14 November 2004; BBC Summary of World Broadcasts). By 2004 this had risen to 74,000 and by 2005 to over 87,000 according to the Ministry of Public Security, with most revolving around local land disputes (Watts 2006). This contrasts with a figure of 10,000 for 1994 according to Zhou Yong Kang, Minister of Public Security (Wang 2006). The article by Thireau and Hua (2003) on workers' expression of grievance through letters and the use of arbitration committees reveals that, in Shenzhen alone, half of all complaints addressed to the 'Letters and Visits Offices' of the municipal government related to labour issues. A newspaper report in 2004 suggests that migrant workers in Shenzhen were increasingly submitting petitions to the local government about excessive overtime, delayed payment of wages and failure to pay social security insurance.[32] Chen (2000) finds that in Henan province, 55.7 per cent of protests concern wage and pension arrears, and 37.7 per cent bankruptcies and mergers.

However, such aggregate figures do not reveal to what extent the protests involved migrant workers, though one could surmise that issues of pension arrears, bankruptcies and mergers apply more to workers made redundant from state enterprises, and wage arrears to migrant workers. Moreover, given that migrant workers form the bulk of the workforce in certain sectors, such as the construction industry, and in certain parts of China where foreign investment is concentrated, such as Guangdong, Shanghai and Fujian, protests in these sectors and areas are likely to involve a significant proportion of migrant workers. Reports from China Labour Bulletin, in local and international newspapers, as well as research by academics reveal an escalation of protests by workers from the late 1980s onwards. Where these involve migrant workers, the key issues have involved delays in payment of wages, excessive overtime, unsafe working conditions and poor canteen and dormitory facilities. To illustrate this, 100 migrant workers blocked roads in Xian City in January 2006, demanding the payment of back wages owed to them so they could return home at the Spring Festival.[33] The demands are thus economic in nature rather than for proper worker representation and the right to form a trade union. However, according to *China Labour Action Express* (2005), in April 2005 migrant workers for the first time organized a strike to form a trade union. Ten thousand workers in Uniden Electronics Factory, a Japanese-invested enterprise in Fuyong town, Shenzhen, went on strike to demand the right to form a trade union branch.[34]

Conclusion

While migrant workers play a central role in fuelling rapid economic growth in China, for a long time they have been treated institutionally and socially as second-class citizens. Since the 16th Party Congress in 2002, central government and

Party leaders, concerned about widening inequalities and rising social unrest, have called for improved protection of migrant workers' rights and instituted various policies, schemes and measures to this end. Though this offers some prospects of improved conditions for migrant workers, much will depend on the proper implementation of such policies and, ultimately, the negotiation of interests among workers, employers and government at the local level. The ability of migrant workers to organize themselves is crucial to the articulation and negotiation of their interests.

However, as demonstrated through the overview presented here, organizing around the interests of migrant workers, whether by others on behalf of migrant workers or by migrant workers themselves, has been limited in China. It is only since the mid-1990s that organizations addressing migrant workers' needs have begun to develop. While these have tended to focus on the legal aspects of migrant workers' rights, few adopted an empowerment agenda that would facilitate migrant workers to organize themselves. Nor did the majority of these organizations seek to advocate on behalf of migrant workers for changes that address the structural sources of inequalities, such as the divisive residence system and the concomitant inequalities concerning access to housing, welfare, medical care and schooling. The services provided by organizations working on behalf of migrant workers are nevertheless valuable, not least because they draw attention to some of the most pressing issues faced by migrants. However, they are also hampered in their effectiveness by the lack of a unified approach and strategy, their small scale and the absence of horizontal coordination between groups. These weaknesses are, in turn, informed by the broader political and institutional environment within which civil society organizations operate in China. Their financial dependence on external donor organizations also raises issues about their sustainability.

More troubling, however, is the absence of organizations initiated and run by migrant workers. This is due in part to the difficulties of organizing around a mobile workforce based on workplace relations but without a common identity, but also mainly because of government restrictions on self-organizing by migrant workers. The government prohibition on the formation of independent migrant workers' associations or trade unions aims to preserve the monopoly of the ACFTU on labour representation and ultimately to protect the regime from perceived political threat. The success of such an approach depends on the availability of alternative channels of interest articulation and participation in policy processes. Though the ACFTU has made some efforts to recruit migrant workers, given the workers' lack of confidence in the effectiveness of the union as well as its structural proximity to the Party, it is unlikely that it can provide a timely and effective route for migrant workers to pursue their diverse interests. Furthermore, citizen participation in policy processes in China is minimal, despite recent efforts to open them up through public hearings and consultation with experts. Though we can expect an increase in government-organized migrant workers' associations, the inherent ambiguity about whose interests such top-down initiatives actually represent, and the tension between control and representation, are likely to constrain free expression by migrant workers. In the absence of alternative channels of expression, we can expect spontaneous protests by migrant workers to continue, if not increase.

Notes

1 See government work report of Wen Jiabao delivered to the National People's Congress in March 2006 (text of Chinese Premier's government work report at NPC session'; China Central TV-1, Beijing, March 5 2006, translated in BBC Monitoring Global Newsline, Asia–Pacific Political, March 7 2006).
2 These figures are, however, contested. Using the internationally recognized poverty line of an average income of US$ 1 per day, there are 90 million people living in poverty in China (Xinhua News Agency, Beijing, 27 June 2005).
3 I am very grateful to the research assistance provided by Diana Lewis, postgraduate researcher at the LSE.
4 For further details about pre-revolutionary and pre-reform civil society, see White *et al.* (1996: 15–23), Ma (1994), Huang (1993) and Rowe (1993).
5 Xue (1988) argued that social organizations could 'serve as a bridge between the state and the enterprises' and thereby a way of indirectly regulating the economy.
6 These were the new Management Regulations on the Registration of Social Organizations, which replaced the 1950 regulations. Under this, social organizations were required to affiliate to a supervisory body (*guakao danwei*), which was responsible for overseeing the day-to-day affairs of the association, and to register with the Division of the Supervision of Social Organizations in the Ministry of Civil Affairs.
7 The first women studies centre to be set up in China was in Zhengzhou University in 1987. By 2001 there were over 40 such centres in various universities and colleges.
8 For example, registering organizations had to have at least 50 individual or 30 institutional members. National level organizations were required to have at least RMB 100,000 for their activities and a fixed location. Apart from the Regulations for the Registration and Administration of Social Organizations, two other sets of regulations were also introduced in 1998; namely, the Provisional Regulations for the Registration and Management of Non-Commercial Institutions and the Provisional Regulations for the Registration and Management of Popular Non-Enterprise Work Units.
9 The number of registered social organizations fell from 181,060 in 1993 to 162,887 at the end of 1998 (Ministry of Civil Affairs 2000: 132). In 1999 alone, the registration of 35,236 social organizations was cancelled. By the end of the year there were only 136,841 registered social organizations in China.
10 An internal circular issued by the Ministry of Civil Affairs in 1989 prohibited national-level social organizations from setting up their own branches at the provincial level, thereby preventing the development of organizations with nationwide links (White *et al.* 1996: 105).
11 These were the Regulations for the Management of Foundations, which took effect on June 1, 2004 and annulled the 1998 Measures for the Management of Foundations; and the China Public Welfare Donation Law, which took effect on September 1999.
12 For a very detailed discussion of the idea of 'small government, big society', see Ru (1998).
13 It should also be noted that this argument may also just be used to justify the monopoly of the All-China Federation of Trade Unions over the representation of labour.
14 This division has been institutionalized from the 1950s onwards through various means such as the urban residence regulations, and the wage system, the system for access to welfare and medical care.

15 See 'Chinese migrant workers from rural areas totalled 113.9 million in 2003', Xinhua News Agency, Beijing, 14 May 2004.

16 See 'Chinese vice-premier urges settlement of back payments for workers' in Xinhua News Agency, Beijing, 23 August 2004. For companies that miss the payment deadline set by the Ministry of Labour and Social Security under the State Council labour regulation on the back payment of wages, such companies will have to pay an extra 50 to 100 per cent of the amount owed to workers. For further details see 'Beijing cracking down on unpaid wages of migrant workers', *South China Morning Post* website, November 16 2004.

17 For further details see 'China capital lifts regulations against immigrants', Xinhua News Agency, Beijing, 26 March 2005.

18 For example in February 2006 a district in Guangyun City, Sichuan province, set up a migrant workers' trade union to protect the rights of migrant workers. Below the district trade union are 18 county migrant workers' trade unions and 148 village branch unions (see *'Guangyuan nongmingong dai 'zheng' shanglu'* (A migrant worker in Guangyuan on the road to waiting for a card) in *Gongren Ribao* (Workers' Daily), 10 Febryary 2006, downloaded from www.grrb.com.cn/news/news in March 2006.

19 In his article Wang points out that the lack of a national leading body to provide overall planning and coordination for the management of migrant workers is an institutional weakness that could hinder the successful implementation of separate measures and policies.

20 I am grateful to the Department for International Development (DfID) which provided me with a small grant to carry out research on organizing around marginalized groups in China from 2000 to 2002. Over 66 interviews were conducted with civil society organizations and government officials, including organizations working in the field of migrant workers. I am also grateful to the DfID for part-funding research on labour organizations between 2002 and 2005.

21 This was initiated in 1993 by Xie Li Hua, who was Chief Editor of the publication *China Women News* produced under the All-China Women's Federation.

22 For example, in 2003 it raised RMB 30,000 through a charity dinner for the fund.

23 By 2005, more than 40 migrant women had received support from the Emergency Relief Fund.

24 Between 1996 and 2000 there were also two employees from the local district trade union, but they reportedly did not play an active role in the activities of the Centre (interview, E37, March 2001). By 2004 the centre had ten mainland and three Hong Kong organizers (Pun and Yang 2004).

25 For example, it facilitated female workers to express their grievances and write a letter to the local Labour Bureau to address a range of issues, such as low wages, overtime and poor living conditions (interview, E37, 2001).

26 Interviews, E39, E45, E57 E58, March 2001.

27 Interview, E45, March 2001.

28 The government continued to fund the salary and welfare protection of the head of the association and also funded the salaries of the joint defence teams (Gu 2005). Moreover, it continues to undertake certain government functions such as running the ten service centres, whilst also seeking to develop a wider range of activities to reach all migrant workers in Haikou city.

29 For example, in its 2002–2005 China Country Strategy Paper, E19, it is stated: 'We will ensure that our work in China addresses the causes of vulnerability as well as absolute poverty. In practice, this means priority will be given to the needs and rights of groups who are excluded from, or harmed by, processes of development and change. Through our project activities we will support increased voice and access to benefits for groups such as ethnic minorities, migrants, the disabled [...]'

30 In March 2005 36 such awards were made to female migrant workers out of a pool of 300 applicants (www.asiafoundaton.org/Lcations/china_may.html, accessed April 10, 2006.
31 Gu (2005) in fact asserts that it was the first Migrant Workers Association to be formed in China, though his own account of the process points to a top-down initiative from an enlightened local government.
32 'Hong Kong paper reports labour disputes rising in China's Shenzhen', *Hong Kong South China Morning Post* website, 14 August 2004, reported in BBC Monitoring Global Newsline, 14 August 2004.
33 'Migrant workers demand back wages, block roads in China's Xian', *Ta Kung Pao* website, Hong Kong, 21 January 2006, reported in BBC Monitoring Global Newsline, 21 January 2006.
34 In fact, the workers had gone on strike the previous year to demand higher salaries and shorter working hours. Since then, the leaders of the strike had been dismissed and the company had reneged on promises to allow the workers to form a union ('Chinese workers stage strike action demanding union rights at Japanese factory', *Hong Kong South China Morning Post* website, 21 April 2005, reported in BBC Monitoring Global Newsline, 21 April 2005).

References

Chan, A. and Unger, J. (1996) 'Corporatism in China. A developmental state in an East Asian context', in McCormick, B.L. and Unger, J. (eds) *China after Socialism. In the Footsteps of Eastern Europe or East Asia?* London: M.E. Sharpe.

Chen F. (2000) 'Subsistence crises, managerial corruption and labour protests in china', *The China Journal*, 44(July): 41–63.

China Brief (2001) '250 Chinese NGOs. A special report from China Brief', August, pp. 1–307.

China Development Brief (2002) 'New moves for Pearl Delta Migrants', China Development Brief, January.

Ding, X.L. (1994) 'Institutional amphibiousness and the transition from communism: The case of China', *British Journal of Political Science*, 24(3): 293–318.

Du, J. (2001) 'Gender and governance: The rise of women's organisations', in Howell, J. (ed.) *Governance in China*. New York: Rowman and Littlefield, pp. 172–192.

Gu, X.L. (2005) 'From management to self-rule – home of migrant workers, model of innovation in Longhua District, Haikou City', 31 January, The World and China Research Institute. Available at: http://www.world-china.org/00/back081.htm, accessed February 2006.

Howell, J. (1997) 'Post-Beijing reflections: creating ripples, but not waves in China', *Women's Studies International Forum*, 20(2): 235–252.

Howell, J. (2004) 'New directions in civil society: Organising around marginalised interests', in Howell, J. (ed.) *Governance in China*. New York: Rowman and Littlefield, pp. 143–171.

Howell, J. (2006) 'New democratic trends in China? Reforming the All-China Federation of Trade Unions', IDS Working Paper 263 (March), pp. 1–28.

Hsiung, P.-C., Jaschok, M. and Milwertz, C. with Red Chan (eds) (2001) *Chinese Women Organising. Cadres, Feminists, Muslims and Queers.* New York: Berg.

Huang, P.C. (1993) 'Public sphere/civil society in China? The third realm between state and society', *Modern China*, 19(2): 216–240.

Ma, S.Y. (1994) 'The Chinese discourse on civil society', *The China Quarterly*, 137(March): 180–193.

Madsen, R. (1993) 'The public sphere, civil society and moral community: A research agenda for contemporary China studies', *Modern China*, 19(2): 183–198.

Ministry of Civil Affairs (2000) *China Civil Affairs' Statistical Yearbook*. Beijing: China Statistics Press.

Murison, S. (2005) 'Evaluation of DFID development assistance: Gender equality and women's empowerment. Phase II thematic evaluation: Migration and development', DFID Working Paper 13 (March), pp. 1–29.

Perrement, M. (2005) 'ILO business start-up programme targets migrants', China Development Brief, November.

Pan, P.P. (2002) 'When workers organise, China's party-run unions resist', Washington Post Foreign Service, 15 October.

Pei, M. (1998) 'Chinese civic associations: An empirical analysis', *Modern China*, 24(3): 285–318.

Pun, N. and Yang, L.M. (2004) 'The Chinese Working Women's Network', Against the Current. Available at www.solidarity-us.org/atc/113Luce.html, accessed March 2006.

Qian, T. (2005) 'Group brings music, rights awareness to migrant workers', China Development Brief, March, pp. 1–2.

Rowe, W. (1993) 'The problem of 'civil society' in late imperial China', *Modern China*, 19(2): 139–157.

Ru, X. (ed.) (1998) *Xiao zhengfu da shehui de lilun yu shiqian* (Theory and Practice of Small Government and Big Society). Beijing: Social Sciences Collections Publishing House.

Solinger, D. (1999) *Contesting Citizenship in Urban China: Peasant Migrants, the State, and the Logic of the Market*. Berkeley: University of California Press.

Sullivan, L. (1990) 'The emergence of civil society in China, spring 1989', in Saich, T. (ed.) *The Chinese People's Movement: Perspectives on Spring 1989*. London: M.E.Sharpe, pp. 126–144.

Tang, R. (2004) 'The winds of change', *Weekend Standard*, 31 July 2004. Available at: http://www.hartford-hwp.com/archives/55/689.html, accessed April 2006.

Thireau, I. and Hua, L. (2003) 'The moral universe of aggrieved Chinese Workers: Workers' appeals to arbitration committees and letters and visits offices', *The China Journal*, 50(July): 83–106.

Unger, J. (1996) 'Bridges in private business, the Chinese Government and the rise of new associations', *The China Quarterly*, 147(September): 795–819.

Wang, I. (2006) 'Social unrest in China', *Hong Kong South China Morning Post*, 20 January.

Wang Y. (2005) 'Full employment for peasants is the key to being free from the dual economic structure', *Renmin Ribao* (People's Daily), 31 October, translated in BBC Monitoring Global Newsline, Asia-Pacific Economic, 1 November 2005.

Wank, D. (1995) 'Private business, bureaucracy and political alliance in a Chinese City', *Australian Journal of Chinese Affairs*, 33(January): 55–71.

Watts, J. (2006) 'Police turn water cannon on rural protest in China', *The Guardian*, 14 April.

White, G., Howell, J. and Shang, X. (1996) *In Search of Civil Society. Market Reform and Social Change in Contemporary China*. Clarendon Press: Oxford.

Xie, C.L. (2002) 'Wailai gong zizhi zuzhi chuxian Zhejiang Ruian (Migrant workers' self-organization first appeared in Ruian, Zhejiang), *Nanfang Zhoumou*, 7 April.

Xue, M. (1988) 'Establish and develop non-governmental self-management organizations in various trades', *Renmin Ribao* (People's Daily), 10 October, translated in Foreign Broadcast Information Service, China Report, 88/201.

Index